ARCO REGENTS REVIEW SERIES

BIOLOGY

Contributing Editors

Louis Bancheri, M.A., M.S.
Chairman, Science Dept.
Sewanhaka High School
Floral Park, NY

Imelda Gallagher
Teacher of Biology
Sewanhaka High School
Floral Park, NY

Carol Mitch
Science Teacher
Great Neck North High School
Great Neck, NY

James D. Stiles
Science Teacher
Binghamton City High School
Binghamton, NY

A NAUTILUS COMMUNICATIONS BOOK

ARCO PUBLISHING, INC.
NEW YORK

Published by Arco Publishing, Inc.
215 Park Avenue South, New York, N.Y. 10003

Library of Congress Cataloging in Publication Data
Main entry under title:

Biology.

    (Arco regents review series)
    "A Nautilus Communications book."
    1. Biology. I. Bancheri, Louis. II. Series.
QH308.2.B5644 1983   574   82-20671
ISBN 0-668-05697-5 (pbk.)

Printed in the United States of America

CONTENTS

Introduction

For the first time, here is a new way to prepare for the Regents Examinations. Now, instead of only taking prior examinations and checking your answers, with an *Arco Regents Review* book you can quickly determine what you already know, and what topics require additional study.

This unique book presents the subject by specific topics. All you have to do is take the Pretest that appears at the beginning of the book. Check your answers against the Topic Analysis Key to evaluate your current strengths and weaknesses. Then, you can study only those topics that require additional work, or if you wish, go through the book systematically, topic by topic.

In addition, each chapter contains valuable review material, so you will have necessary information at your fingertips. And, as you will see, all of the questions are answered, and where appropriate, complete explanations of those answers are presented. Mark your progress, and be sure to read the explanations carefully to understand why an answer is correct or not.

By paying attention, you will be able to develop important test-taking skills. These include:

1. Reading directions carefully.
2. Understanding the question.
3. The process of elimination in multiple-choice questions.
4. Identifying specifics, for questions and essays.

When you have completed the appropriate chapters, you can then evaluate how much progress you have made by taking the Posttest, checking your answers, and marking the Topic Analysis Key. If you need more review, go back to the appropriate chapter, or ask your teacher for help.

Finally, take a recent N.Y. State Regents Examination, so you will be mentally, as well as academically, prepared for the actual exam. Again, the Topic Analysis Key will give you one more chance to evaluate your strengths and weaknesses.

We think you will be pleased with this book, and will find that your preparation for the N.Y. State Regents will become faster, easier, and more systematic. More important, however, your studying will be more effective, and hopefully will result in higher Regents scores.

NOTE: Many of the questions in this book may appear similar within a chapter. In fact, the answers may, indeed, be the same. However, it is important to be aware of the many different ways the same, or similar, questions may be asked on an examination, in order to recognize them, and answer them correctly.

Pretest

Before beginning your review of the various topics, we suggest you take this Pretest. It is a broad overview of the subject and will help you evaluate what you already know, and where you need additional review. Think of it as a time saver. By taking the Pretest you can avoid unnecessary review, and give yourself more time to concentrate on those topics that require more of your time and study.

When you have finished the test and checked your answers, consult the *Topic Analysis Key*. It will help you determine how and where to spend your study time.

Answer all of the questions in the Pretest. Work steadily as you would on a regular examination.

Directions: Each question is followed by four choices. Underline the correct choice.

1. Which lists the nervous-system structures in order of increasing size?
   (1) neuron, nerve, ganglion, plexus
   (2) nerve, ganglion, neuron, plexus
   (3) neuron, plexus, ganglion, nerve
   (4) ganglion, plexus, nerve, neuron

2. In pea plants, the trait for smooth seeds is dominant over the trait for wrinkled seeds. When two hybrids are crossed, which results are most probable?
   (1) 75% smooth and 25% wrinkled seeds
   (2) 100% smooth seeds
   (3) 50% smooth and 50% wrinkled seeds
   (4) 100% wrinkled seeds

3. Which structure is formed as a result of the process of cleavage?
   - (1) egg cell
   - (2) sperm cell
   - (3) zygote
   - (4) blastula

4. The process by which excess water leaves a plant through the stomates is known as
   - (1) locomotion
   - (2) photosynthesis
   - (3) hydrolysis
   - (4) transpiration

5. In which organ of the human body is urea produced as a result of the breakdown of amino acids?
   - (1) liver
   - (2) pancreas
   - (3) stomach
   - (4) small intestine

6. Amino acids derived from the digestion of a piece of meat are transported to living cells of an animal. In the cell they are
   - (1) converted to cellulose
   - (2) used to attack invading bacteria
   - (3) synthesized into specific proteins
   - (4) incorporated into glycogen molecules

7. Which of the following variables has the *least* direct effect on the rate of a hydrolytic reaction regulated by enzymes?
   - (1) temperature
   - (2) pH
   - (3) carbon dioxide concentration
   - (4) enzyme concentration

8. In pea plants, green pods are dominant over yellow pods. In which of the following groups would members have the same phenotype but different genotypes?
   - (1) pure green pods and homozygous green pods
   - (2) hybrid green pods and homozygous green pods
   - (3) yellow pods and hybrid green pods
   - (4) yellow pods and homozygous yellow pods

9. In sorghum plants, red stem is dominant over green stem. If 1,000 seeds from a sorghum plant germinated to produce 760 red plants and 240 green plants, it would be most reasonable to assume that the parental genotypes were
   - (1) $Rr \times Rr$
   - (2) $RR \times rr$
   - (3) $Rr \times RR$
   - (4) $Rr \times rr$

10. The classification group which shows the greatest similarity among its members is the
    - (1) phylum
    - (2) kingdom
    - (3) genus
    - (4) species

11. Since the publication of Darwin's theory, evolutionists have developed the concept that
    (1) a species produces more offspring than can possibly survive
    (2) the individuals that survive are those best fitted to the environment
    (3) through time, favorable variations are retained in a species
    (4) mutations are partially responsible for the variations within a species

12. All of the members of a single species inhabiting a given location are known as
    (1) an ecosystem
    (2) a population
    (3) a niche
    (4) a food chain

13. Which instrument is used to collect large samples of mitochondria?
    (1) an ultracentrifuge
    (2) an electron microscope
    (3) a phase contrast microscope
    (4) a microdissecting needle

14. Which world biome has the greatest numbers of organisms?
    (1) tundra
    (2) tropical forest
    (3) temperate deciduous forest
    (4) marine

15. Most of the minerals within an ecosystem are recycled and returned to the environment by the direct activities of organisms known as
    (1) producers
    (2) secondary consumers
    (3) decomposers
    (4) primary consumers

16. Which would *not* be essential in a self-sustaining ecosystem?
    (1) a constant source of energy
    (2) living systems capable of incorporating energy into organic compounds
    (3) equal numbers of plants and animals
    (4) a cycling of materials between organisms and their environment

17. According to the heterotroph hypothesis, where were the first cell-like structures probably formed?
    (1) in fresh water
    (2) on land
    (3) in the ocean
    (4) within crustal rocks

18. A person who is homozygous for blood type A has a genotype which may be represented as
    (1) $I^a I^b$
    (2) $I^a I^a$
    (3) $I^a i$
    (4) ii

19. For a given trait, the two genes of an allelic pair are not alike. An individual possessing this gene combination is said to be
    (1) homozygous for that trait
    (2) heterozygous for that trait
    (3) recessive for that trait
    (4) pure for that trait

20. Which structures code information for the inheritance of traits?
    (1) nuclear membranes   (3) vacuoles
    (2) cell membranes   (4) genes

21. The male reproductive organ of a flowering plant is the
    (1) pistil   (3) stigma
    (2) ovary   (4) stamen

22. In human skin cells, the products of a normal mitotic cell division are
    (1) 4 diploid cells   (3) 2 monoploid cells
    (2) 2 diploid cells   (4) 4 monoploid cells

23. The development of an unfertilized egg is known as
    (1) parthenogenesis   (3) germination
    (2) symbiosis   (4) synapsis

24. Carbon dioxide enters the moist internal surface of a leaf through
    (1) stomates   (3) phloem cells
    (2) lenticels   (4) root hairs

25. In most green plants, which wavelengths of light are most effective in the conversion of radiant energy into the chemical energy of organic compounds?
    (1) red and green   (3) yellow and green
    (2) yellow and blue   (4) red and blue

26. Most toxic products of plant metabolism are stored in the
    (1) lenticels   (3) stomates
    (2) vacuoles   (4) chloroplasts

27. Which is an example of mechanical digestion?
    (1) a rabbit chewing grass
    (2) an earthworm hydrolyzing protein
    (3) a Hydra performing intracellular digestion
    (4) a contractile vacuole functioning in a Paramecium

28. Which organism uses chitinous appendages for locomotion?
    (1) grasshopper   (3) Paramecium
    (2) Hydra   (4) human

29. Which life function is primarily involved in the conversion of the energy stored in organic molecules to a form directly usable by a cell?
    (1) absorption   (3) digestion
    (2) circulation   (4) respiration

30. Seed plants generally absorb nitrogen from the soil in the form of
    (1) urea   (3) proteins
    (2) nitrates   (4) amino acids

31. Members of the protist kingdom include
    (1) sponges, bacteria, and algae
    (2) protozoa, bacteria, and fungi
    (3) algae, bryophytes, and fungi
    (4) sponges, bryophytes, and protozoa

32. According to the heterotroph hypothesis, which substance was missing
    from the environment of the Earth prior to the origin of life?
    (1) ammonia molecules   (3) hydrogen molecules
    (2) methane molecules   (4) oxygen molecules

33. Based on modern evolutionary theory, the development of a new species
    would most likely be associated with
    (1) a constant environment   (3) geographic isolation
    (2) stable gene pools   (4) a lack of mutations

34. Fossils are used as evidence for evolution because they
    (1) may show a pattern of consecutive changes
    (2) are as old as the oldest rocks on the Earth
    (3) are composed of mineral substances
    (4) may be formed in lava rock

35. An inherited metabolic disorder known as phenylketonuria (PKU) is char-
    acterized by severe mental retardation. This condition results from the
    inability to synthesize a single
    (1) enzyme   (3) vitamin
    (2) hormone   (4) carbohydrate

36. Of the 500 eggs produced by a certain female frog, only 10% developed
    into adult frogs. Which part of Darwin's theory does this best illustrate?
    (1) Favorable variations are not inherited.
    (2) There is a struggle for existence among organisms.
    (3) Mutations occur by chance.
    (4) Mating occurs in a random manner in a species.

37. Which occurs in a plant cell but *not* in an animal cell during mitotic cell
    division?
    (1) formation of spindle fibers   (3) formation of a cell plate
    (2) chromosome duplication   (4) cytoplasmic division

38. Which is the best explanation for the growth pattern of the plant represented in the diagram below?

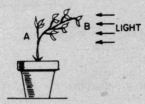

(1) Light affects the distribution of auxins.
(2) More photosynthesis takes place on side *B* than on side *A*.
(3) Cells on side *B* grow faster than those on side *A*.
(4) Light stimulates a geotropic response.

39. Certain bacteria obtain energy for carbon fixation by converting inorganic nitrates to nitrogen gas. Bacteria which have this pattern of nutrition are known as
(1) autotrophs   (3) omnivores
(2) heterotrophs   (4) parasites

40. Which activity is an example of intracellular digestion?
(1) a grasshopper chewing blades of grass
(2) a maple tree converting starch to sugar in its roots
(3) an earthworm digesting proteins in its intestine
(4) a fungus digesting dead leaves

41. The movement of materials throughout vascular plants is similar to circulation in vertebrates because both have
(1) specialized transport tissues
(2) veins made of xylem and phloem
(3) arteries, veins, and capillaries
(4) an open transport system

42. Which substance aids in the transmission of nerve impulses across a synapse?
(1) auxins   (3) gastric fluid
(2) neurohumors   (4) lachrymal fluid

43. Based on their respiration and nutrition, humans are usually classified as
(1) aerobic autotrophs   (3) anaerobic autotrophs
(2) aerobic heterotrophs   (4) anaerobic heterotrophs

44. The net flow of materials through the membrane of a cell against a concentration gradient is known as

(1) active transport
(2) passive transport

(3) circulation
(4) transpiration

Directions: Base your answers to questions 45 through 47 on the information below and on your knowledge of biology.

A yeast culture is added to a solution of glucose and water in the absence of oxygen. One hour later, bubbles of carbon dioxide gas are being produced.

45. The process that produces the bubbles of CO_2 is most likely
    (1) carbon fixation
    (2) dehydration synthesis
    (3) anaerobic respiration
    (4) enzyme deactivation

46. For the production of CO_2, which compound might best be substituted for glucose?
    (1) ATPase
    (2) sucrose
    (3) lactic acid
    (4) alanine

47. Which compound in the solution increases in concentration as the carbon dioxide is produced?
    (1) starch
    (2) ethyl alcohol
    (3) glucose
    (4) deoxyribose

48. The process by which digestive enzymes catalyze the breakdown of larger molecules to smaller molecules with the addition of water is known as
    (1) synthesis
    (2) pinocytosis
    (3) hydrolysis
    (4) photosynthesis

Directions (49–53): For *each* phrase in questions 49 through 53, select the biome, *chosen from the list below,* that is best described by that phrase. [A number may be used more than once or not at all.]

Biomes

(1) Taiga
(2) Tropical rain forest
(3) Temperate deciduous forest
(4) Tundra
(5) Desert

49. Has a forest of spruce, fir, and pine trees as climax vegetation

50. Has oak, maple, and beech in its climax forest

51. Has the lowest average temperature

52. Characterized by trees and shrubs which shed their leaves after the summer growing season

53. Has a great daily temperature fluctuation and contains mainly cold-blooded animals which are active at night and many producers which lack broad leaves

54. Which type of evolutionary evidence is represented by these diagrams?

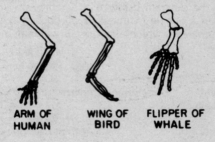

ARM OF HUMAN   WING OF BIRD   FLIPPER OF WHALE

    (1) homologous structures   (3) biochemical similarities
    (2) physiological likenesses   (4) geographic distribution

55. Certain strains of bacteria that were susceptible to penicillin in the past have now become resistant. The probable explanation for this is that
    (1) the mutation rate must have increased naturally
    (2) the strains have become resistant because they needed to do so for survival
    (3) a mutation was retained and passed on to succeeding generations because it had high survival value
    (4) the principal forces influencing the pattern of survival in a population are isolation and mating

56. The principles of dominance, segregation, and independent assortment were first described by
    (1) Darwin   (3) Lamarck
    (2) Watson and Crick   (4) Mendel

57. In guinea pigs, black coat color is dominant over white. If a heterozygous black-coated guinea pig is mated with a white-coated guinea pig, how many different phenotypes with respect to coat color could be expected in the offspring?
    (1) 1
    (2) 2
    (3) 3
    (4) 4

58. The presence of only one X-chromosome in each body cell of a human female produces a condition known as Turner's syndrome. This condition most probably results from the process known as
    (1) polyploidy
    (2) crossing-over
    (3) nondisjunction
    (4) hybridization

Directions: Base your answers to questions 59 through 63 on the diagram below which represents structures found in a female mammal's reproductive system and some processes which might occur within that system.

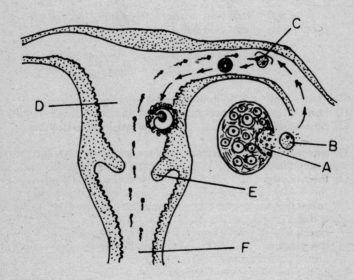

59. The structure labeled *B* represents
    (1) an egg
    (2) a zygote
    (3) a follicle
    (4) an ovary

60. In which structure(s) would sperm cells normally be deposited?
    (1) A, only
    (2) C, D, and E
    (3) C and D, only
    (4) F, only

61. Which process is represented at C?
    (1) fertilization
    (2) cleavage
    (3) implantation
    (4) menstruation

62. The process by which the structure labeled B is released from the structure labeled A is known as
    (1) differentiation
    (2) ovulation
    (3) gastrulation
    (4) germination

63. In which structure will a fetus normally develop?
    (1) E
    (2) F
    (3) C
    (4) D

Directions: Base your answers to questions 64 and 65 on the word equation below.

glucose → 2 pyruvic acid → 2 ethyl alcohol + 2 carbon dioxide + energy

64. Organisms which carry on this process include
    (1) Hydrae
    (2) earthworms
    (3) yeasts
    (4) grasshoppers

65. The process represented by the word equation is known as
    (1) aerobic respiration
    (2) fermentation
    (3) chemosynthesis
    (4) dehydration synthesis

66. Locomotion is accomplished by the interaction of muscles and chitinous appendages in the
    (1) Hydra
    (2) Paramecium
    (3) grasshopper
    (4) human

67. Nephridia function in the excretion of
    (1) carbon dioxide and glycerol
    (2) ammonia and urea
    (3) feces and bile salts
    (4) amino acids and nitrogen

Directions: Base your answers to questions 68 through 70 on the diagram below which represents a small fragment of a protein molecule.

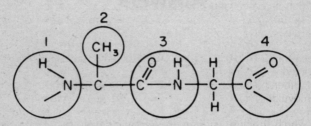

68. How many amino acids are bonded together to form the fragment represented by the diagram?
    (1) 5   (3) 3
    (2) 2   (4) 4

69. Which circled area represents a peptide bond?
    (1) 1   (3) 3
    (2) 2   (4) 4

70. Which circled area represents a group that may vary from molecule to molecule and therefore provides for a variety of different amino acids?
    (1) 1   (3) 3
    (2) 2   (4) 4

Answers

1. 1	16. 3	31. 2	46. 2	61. 1
2. 1	17. 3	32. 4	47. 2	62. 2
3. 4	18. 2	33. 3	48. 3	63. 4
4. 4	19. 2	34. 1	49. 1	64. 3
5. 1	20. 4	35. 1	50. 3	65. 2
6. 3	21. 4	36. 2	51. 4	66. 3
7. 3	22. 2	37. 3	52. 3	67. 2
8. 2	23. 1	38. 1	53. 5	68. 2
9. 1	24. 1	39. 1	54. 1	69. 3
10. 4	25. 4	40. 2	55. 3	70. 2
11. 4	26. 2	41. 1	56. 4	
12. 2	27. 1	42. 2	57. 2	
13. 1	28. 1	43. 2	58. 3	
14. 4	29. 4	44. 1	59. 1	
15. 3	30. 2	45. 3	60. 4	

Explanatory Answers

1. **(1)** Neurons are nerve cells which form nerves. Nerves form ganglions and ganglions form a plexus. INCORRECT CHOICES: (2), (3), and (4) These choices do not follow the sequence described above—from the smallest units to larger ones.

2. **(1)** A 3:1 ratio of dominant to recessive results from the cross of two hybrids. INCORRECT CHOICES: (2) This will result from crossing a pure dominant plant with any other plant. (3) This results from crossing a pure recessive with a hybrid plant. (4) This results from crossing two recessive plants.

3. **(4)** The hollow ball of cells called the blastula forms during the process of cell division known as cleavage. INCORRECT CHOICES: (1) The egg cell is the female sex cell. (2) The sperm cell is the male gamete. (3) The zygote or fertilized egg divides without increasing in size during cleavage; it is formed as a result of the union of the egg and sperm cell.

4. **(4)** Transpiration is the loss of excess water through openings in the leaves. INCORRECT CHOICES: (1) Locomotion is the movement of organisms from place to place. (2) Photosynthesis is the process by which green plants make food. (3) Hydrolysis is the chemical breakdown of large molecules into smaller ones.

5. **(1)** The liver changes nitrogenous wastes to less toxic urea. INCORRECT CHOICES: (2) The pancreas secretes digestive juices and hormones. (3) The stomach digests proteins. (4) The small intestine is the principal organ for digestion and absorption.

6. **(3)** Proteins are the products of amino acid synthesis. INCORRECT CHOICES: (1) Cellulose is a polysaccharide made from sugars, not amino acids. (2) White blood cells are used to attack bacteria. (4) Glycogen is produced to store sugars, not amino acids.

7. **(3)** Carbon dioxide concentration is not a major factor in enzyme controlled reactions. INCORRECT CHOICES: (1) In an enzyme controlled reaction, the rate increases when the temperature increases from 0°C to about 40°C, and decreases at higher temperatures. (2) Individual enzymes are effective at specific pH environments. (4) To a point, an increase in the enzyme concentration increases the rate of reaction.

8. **(2)** The hybrid and homozygous green pods have the same phenotype or appearance but a different genetic makeup or genotype. INCORRECT CHOICES: (1) Pure green pods are homozygous. (3) Yellow pods and hybrid green pods have a different appearance. (4) All yellow pods must have the homozygous genotype.

9. **(1)** A cross between Rr and Rr will result in a 3:1 ratio of red plants to green plants. INCORRECT CHOICES: (2) $RR \times rr$ would produce all Rr or hybrid red plants. (3) $Rr \times RR$ would result in all red plants, 50% pure and 50% hybrid. (4) $Rr \times rr$ would produce 50% red (Rr) and 50% green (rr) plants.

10. **(4)** A species is the smallest subdivision and includes organisms which show the greatest similarity. INCORRECT CHOICES: (1) A phylum is a large classification division under kingdoms. (2) All organisms are divided into one of three major kingdoms, Plants, Animals, and Protists. (3) A genus is a group of closely related species.

11. **(4)** The occurrences of mutations were observed in nature and in the genetics laboratory many years after Darwin postulated his theories. INCORRECT CHOICES: (1), (2), and (3) These evolutionary theories or principles

were observed and proposed by Darwin in his publication on natural selection.

12. **(2)** A population includes all the organisms of a species in an area. INCORRECT CHOICES: (1) No single species but all living and nonliving things function together in an ecosystem. (3) A niche is the role that a single species plays in the community. (4) A food chain describes the transfer of food energy through a series of organisms.

13. **(1)** The ultracentrifuge can separate mitochondria from other organelles in the cells because of their different densities. INCORRECT CHOICES: (2) The electron microscope can make visible details in the fine structure of mitochondria and other cell organelles. (3) The phase contrast microscope can make visible unstained organelles in living cells. (4) With a micro-dissecting needle, it is possible to remove individual large organelles from a cell.

14. **(4)** The two-thirds of the earth's surface which is covered by water, contains more plants and animals than there are on land. INCORRECT CHOICES: (1) The tundra zone which surrounds the Arctic ocean is permanently frozen to a depth of several meters and contains a relatively small number of living organisms. (2) The tropical or rain forest (or jungle) has many plants and animals. The high vegetation is dense and little light reaches the forest floor. It occupies less of the earth's surface than the marine biome. (3) The temperate deciduous forest has a moderate, well-distributed rainfall, but less area than the marine biome.

15. **(3)** Decomposers act on dead organic material to break down the complex organics into simple substances thereby returning them to the soil. INCORRECT CHOICES: (1) Producers are photosynthetic and derive their nutrition from the Sun. (2) Secondary consumers rely on primary consumers for food. (4) Primary consumers feed on the producers.

16. **(3)** In an ecosystem there must be more plants to produce food for the animals. INCORRECT CHOICES: (1), (2), and (4) All self-sustaining ecosystems require the input of a constant supply of energy, organisms which produce food, and a constant recycling of materials.

17. **(3)** The aggregates or coacervates formed in the hot thin soup had around them an electrically charged membrane which gave this structure the assets of a cell. INCORRECT CHOICES: (1) At the point of formation fresh water wasn't to be found. (2) The land surfaces were not suited to the formation of primitive cells because both these cells and modern cells need water as a factor of survival. (4) At the time of formation crustal rocks were being formed and were too hot to sustain life.

18. **(2)** Both these represent identical alleles for homozygous type *A* blood. INCORRECT CHOICES: (1) This represents alleles for *AB* type blood. (3) This represents the alleles for heterozygous *A* type blood. (4) This represents the alleles for *O* type blood.

19. **(2)** Two genes for a trait make up an allelic pair. If the genes are dissimilar, the individual is said to be heterozygous for the trait. INCORRECT CHOICES: (1) If an individual is homozygous, both genes of an allelic pair are the same. (3), (4) Genes, not individuals, are said to be recessive or pure for a trait.

20. **(4)** Genes contain the DNA code for the inheritance of traits. INCORRECT CHOICES: (1) The nuclear membrane surrounds the nucleus which contains the genes. (2) The cell membrane is the outer boundary of all cells. (3) Vacuoles are areas in the cytoplasm specialized for storage, not coding traits.

21. **(4)** The stamen or male reproductive organ of a flowering plant is composed of the anther sac with pollen and the filament. INCORRECT CHOICES: (1) The pistil is the entire female reproductive organ of a flowering plant. (2) The ovary is the part of the pistil in which the ovules or eggs are formed. (3) The stigma is the part of the pistil to which the pollen grains become stuck during pollination.

22. **(2)** Two cells with two chromosomes of each pair (diploid number) are produced by mitosis. INCORRECT CHOICES: (1) Two not four diploid cells result from mitotic cell division. (3) Monoploid cells are the products of meiotic not mitotic cell division. (4) Four monoploid cells normally result from reduction division or meiosis.

23. **(1)** Parthenogenesis takes place when an egg develops without being fertilized by a sperm cell. INCORRECT CHOICES: (2) Symbiosis is a relationship in which two organisms of different species exist together. (3) Germination is the development of a seed into a mature plant. (4) Synapsis is the pairing of homologous chromosomes during meiosis.

24. **(1)** The moist internal surfaces of leaves is the site for several chemical reactions and processes. The route to those sites is provided by pores called stomates. INCORRECT CHOICES: (2) Lenticels allow gas exchange on stem surfaces not leaf surfaces. (3) Phloem cells specialize in the translocation of food not the passage of gases. (4) Root hairs increase surface area for the absorption of water in roots.

25. **(4)** The question is asking which wavelengths of light do chlorophyll molecules react to best. The answer simply is red and blue light. The

other colors are reflected to some degree, more so than red and blue. INCORRECT CHOICES: (1) Red is the right color but its complement of green is totally reflected. (2) Yellow is a close wavelength to green and would be somewhat reflected. (3) Both of these colors are reflected by the chlorophyll molecules.

26. **(2)** Vacuoles are organelles adapted for storage of water, food, and toxic wastes. INCORRECT CHOICES: (1) Lenticels are openings in the stems of plants. (3) Stomates are openings in the leaves of plants adapted for gas exchange. (4) Chloroplasts are the sites for photosynthesis.

27. **(1)** Mechanical digestion involves the breakdown of food materials to small particles through purely physical means such as chewing. INCORRECT CHOICES: (2) Hydrolyzing protein leads to its chemical breakdown to amino acids. (3) Intracellular digestion refers to enzymatic hydrolysis inside the cells of the organism. (4) The contractile vacuole functions in the excretion of excess water from a Paramecium.

28. **(1)** Locomotion in the grasshopper is accomplished by the interaction of muscles with chitinous appendages (legs and wings). INCORRECT CHOICES: (2) The Hydra is essentially sessile but does have contractile fibers which permit some motion. (3) Paramecium uses cilia for locomotion. (4) Locomotion in humans is accomplished by the interaction of muscles which move the bones of the skeleton.

29. **(4)** Respiration converts the energy stored in organic compounds, such as glucose, into ATP; this form of energy is readily available for use by the cell. INCORRECT CHOICES: (1) Absorption is the taking in of materials by a cell or an organism. (2) Circulation is the distribution of materials within an organism. (3) Digestion is breaking down large, insoluable molecules into small, soluble ones.

30. **(2)** Nitrogen in the form of nitrates is absorbed by plants for use in protein synthesis. INCORRECT CHOICES: (1) Urea is manufactured by animals to excrete nitrogenous wastes. (3) and (4) Proteins and amino acids must be broken down before absorption by plants.

31. **(2)** In one modern system of classification, the protist kingdom includes protozoa, bacteria, fungi, and algae. INCORRECT CHOICES: (1) Sponges are included in the animal kingdom. (3) Bryophytes are included in the plant kingdom. (4) Sponges are included in the animal, and bryophytes in the plant kingdom.

32. **(4)** Free molecular oxygen is thought to have been missing from the atmosphere prior to the origin of life, in part because at present, oxygen is a result of the photosynthetic activity of living organisms. INCORRECT CHOICES: (1), (2), and (3) According to the heterotroph hypothesis, the primitive oceans contained dissolved ammonia, methane, and hydrogen.

33. **(3)** Isolation causes a shift in gene frequency which results in increased differences within a population. INCORRECT CHOICES: (1), (2), and (4) A constant gene pool and a lack of mutation favor stability and lack of change. Under these conditions, new species are unlikely to evolve.

34. **(1)** Fossils buried in various layers of the earth show, if undisturbed, a chronological order of events. These events may include what organisms lived at a particular time, earth conditions, and possible reasons for extinction. INCORRECT CHOICES: (2) Rocks would tell a scientist nothing about the evolution of a particular species. They may be used to date the age of the species. (3) The normal course of events in fossil formation would be that the soft fleshy tissues decay and the rigid bone structures, if any, may be mineralized. So mineralization would tell us nothing about the process of evolution. (4) Fossils may be formed in any layer of rock, particularly in sedimentary rock. Hot lava would necessarily incinerate the organism leaving virtually no trace of its existence.

35. **(1)** Metabolism requires the proper functioning and synthesis of enzymes. PKU is characterized by the individual not having the necessary gene to synthesize tryosine from phenylalanine. INCORRECT CHOICES: (2) A hormone is secreted by an endocrine gland and is not an enzyme. (3) Vitamins are essential for proper body metabolism of food. (4) A carbohydrate is an energy source for the body.

36. **(2)** Darwin's theory included the idea that since more offspring are produced than possibly survive, there is a struggle for existence among organisms. INCORRECT CHOICES: (1) Darwin theorized that those individuals which survive transmit their favorable variations to their offspring. (3) Darwin, like most biologists of his time, did not understand the genetic basis for evolutionary change. (4) Darwin's theory of natural selection involves nonrandom mating in a species.

37. **(3)** In plant cell mitosis, a cell plate is formed to divide the cytoplasm. INCORRECT CHOICES: (1) Spindle fibers are formed during mitosis in both plant and animal cells. (2) The making of duplicate chromosomes occurs in all forms of cell division. (4) Different forms of cytoplasmic division occur in plants and animals.

38. **(1)** We see the plant bending toward the light in this picture. The best explanation for this type of plant behavior is that light destroys the auxins on the surface on which the light is falling. Therefore, all growth on this side stops. Growth on the shaded side does not stop and we see the characteristic bending. INCORRECT CHOICES: (2) Photosynthesis occurs evenly on all plant surfaces provided that there is nearly the same chlorophyll concentration. (3) Auxins are fairly general in their promotion of growth in all meristematic tissue. (4) A geotrophic response is initiated by gravity, not light.

39. **(1)** Autotrophs are organisms which can synthesize organic compounds; in this case the energy from the breakdown of nitrates instead of light energy is used. INCORRECT CHOICES: (2) Heterotrophs cannot make their own food and must take in preformed organic compounds. (3) Omnivores are consumers which eat both plants and animals. (4) Parasites live on or in a host from which they take their food.

40. **(2)** Conversion of starch to sugar in maple tree roots takes place in the cells of the roots for the eventual release of energy. INCORRECT CHOICES: (1) Chewing is a process which does not occur inside a cell. (3) Protein digestion occurs in this instance in the digestive tube, not in the cells. (4) Dead leaves are digested by enzymes secreted from the fungi.

41. **(1)** Specialized transport tissues exist in both vascular plants and verte-brates due to the advancement that these organisms show over other organisms. INCORRECT CHOICES: (2) Vertebrates do not possess veins made of xylem and phloem but rather veins composed of smooth muscle tissue. (3) A plant's vascular system does not include arteries, veins, and cap-illaries but rather xylem and phloem. (4) Neither possesses an open system in that the open system shows relatively little advancement with the exception of the dorsal tubular heart.

42. **(2)** Neurohumors are chemicals secreted by the terminal branches of neurons; they transmit impulses across synapses. INCORRECT CHOICES: (1) Auxins are plant hormones which control growth. (3) Gastric fluid, se-creted by the stomach, contains substances necessary for the digestion of proteins. (4) Lachrymal fluid is produced by tear glands to protect and lubricate the eye.

43. **(2)** Humans use oxygen to release energy from the nutrients they take in; they are aerobic heterotrophs. INCORRECT CHOICES: (1) Aerobic autotrophs use oxygen to release energy from the food they make. (3) Anaerobic autotrophs do not need oxygen for respiration; they make their own food. (4) Anaerobic heterotrophs do not use oxygen to release energy from the food they take in.

44. **(1)** Active transport takes place when materials move from an area of lower concentration to an area of higher concentration. INCORRECT CHOICES: (1) In passive transport materials move from a region of higher concentration to one of lower concentration. (3) Circulation is the movement of materials within a system. (4) Transpiration is the loss of water from leaves.

45. **(3)** Fermentation takes place in yeast in the absence of oxygen; the products are carbon dioxide and alcohol. INCORRECT CHOICES: (1) Carbon fixation is the dark reaction of the process of photosynthesis. (2) Dehydration synthesis is the synthesis of more complex substances from simple ones. (4) Enzyme deactivation is the blocking of the chemical reactions occurring in the presences of enzymes by effectively blocking the enzyme activity.

46. **(2)** Sucrose is a disaccharide composed of glucose and another monosaccharide, fructose. INCORRECT CHOICES: (1) ATPase is the enzyme which catalyzes the reactions involving the production of ATP. (3) Lactic acid is produced in fermentation. (4) Alanine is an amino acid.

47. **(2)** In reading the passage, we find yeast cells fermenting glucose. The net result of this reaction is the production of carbon dioxide and ethyl alcohol. It would stand to reason that as more carbon dioxide is produced the levels of ethyl alcohol would increase. INCORRECT CHOICES: (1) Starch is not a by-product of fermentation. (3) Glucose was the reactant and not the product. (4) Deoxyribose is a sugar found in DNA.

48. **(3)** Hydrolysis is the process by which large organic molecules are broken down into their building blocks with the addition of water molecules. Hydrolysis is another term for chemical digestion. INCORRECT CHOICES: (1) Synthesis is the process by which large organic molecules are built up from smaller ones. (2) Pinocytosis is a process by which molecules too large to pass through the cell membrane may be engulfed and brought within the cell. (4) Photosynthesis is a process by which green plants convert light energy into the chemical bond energy of organic compounds.

49. **(1)** Taiga is a biome characterized by forests of spruce, fir, and pine trees. INCORRECT CHOICES: (2) Tropical rain forest is found in regions with high average temperature and high rainfall. (3) Temperate deciduous forest is characterized by trees and shrubs which shed their leaves after the summer growing season. (4) Tundra has the lowest average temperature and is found north of the taiga. (5) Desert is characterized by a great daily temperature fluctuation and low average rainfall.

50. **(3)** Climax forests of oak, maple, and beech are temperate deciduous forests. INCORRECT CHOICES: (1) Taiga has climax vegetation of spruce, fir, and pine trees. (2) Tropical rain forest has little seasonal temperature variation and lush vegetation with many different species. (4) Tundra is characterized by climax vegetation of lichen and moss. (5) Desert typically has many producers which lack broad leaves.

51. **(4)** Tundra is found at very high altitudes and north of the Arctic Circle and has the lowest average temperature. INCORRECT CHOICES: (1) Taiga has a forest of spruce, fir, and pine trees as climax vegetation. (2) Tropical rain forest has high average temperature and little seasonal temperature variation. (3) Temperate deciduous forest has oak, maple, and beech in its climax forest. (5) Desert has a great daily temperature fluctuation and low average rainfall.

52. **(3)** Temperate deciduous forest is characterized by trees (such as oak, maple, and beech) and shrubs which shed their leaves after the summer growing season. INCORRECT CHOICES: (1) Taiga is characterized by evergreens such as spruce, fir, and pine trees. (2) Tropical rain forest has little seasonal temperature variation and lush vegetation with many different species. (4) No trees grow in the Tundra. (5) Many producers in the desert lack broad leaves.

53. **(5)** The desert has a great daily temperature fluctuation and contains mainly cold-blooded animals which are active at night and many producers which lack broad leaves. INCORRECT CHOICES: (1) Taiga has a climax forest of spruce, fir, and pine trees. (2) Tropical rain forest has little seasonal temperature variation and lush vegetation with many different species. (3) Temperate deciduous forest is characterized by trees and shrubs which shed their leaves after the summer growing season. (4) Tundra has the lowest average temperature.

54. **(1)** Homologous structures are those whose anatomical features are similar: The bones of the arm, the wing, and the flipper are basically similar. INCORRECT CHOICES: (2) Physiological likenesses involve similarities in the functioning of different organisms. (3) Biochemical similarities involve likenesses at the molecular level. (4) Geographic distribution refers to the geographic ranges of different organisms.

55. **(3)** A spontaneous mutation gave certain strains of bacteria resistance to penicillin; this trait had a high survival value and was passed on to succeeding generations. INCORRECT CHOICES: (1) The antibiotic did not increase the mutation rate but acted as a selecting agent. (2) Mutations

do not occur in response to need. (4) Isolation and mating are not the principal forces influencing the survival of these bacteria.

56. **(4)** Mendel researched and developed the theory which described the basic patterns of inheritance. INCORRECT CHOICES: (1) Darwin proposed the theory of natural selection. (2) Watson and Crick developed a model of the DNA molecule. (3) Lamarck proposed the theory of inheritance of acquired characteristics, or use and disuse.

57. **(2)** The matings indicated would result in the appearance of a black-and-white guinea pig. There would then be two different phenotype color. INCORRECT CHOICES: (1) To have one phenotype color appear the cross would have to be modified to something like homozygous black crossed with homozygous white. (3) To see the appearance of three colors, we would need a mating in incomplete dominance. (4) If only two colors originated in the mating, there couldn't possibly be four resulting.

58. **(3)** Nondisjunction is the improper segregation of chromosomes during meiosis. Often this is produced by the failure of chromosomes to properly migrate to the polar regions during the first or second meiotic division and is called anaphase lag. The end result is that one daughter cell has one extra chromosome and the other is lacking one chromosome as in Turner's syndrome. INCORRECT CHOICES: (1) Polyploidy is the condition in which the cell has twice the number of chromosomes. The question is referring to a deletion not an addition. (2) Crossing-over is the exchange of genetic information during meiosis. (4) Hybridization is the crossing of different strains to achieve new genetic characteristics.

59. **(1)** Ovulation releases an unfertilized egg from the ovary. INCORRECT CHOICES: (2) Fertilization, which produces the zygote, occurs in the oviduct. (3) A follicle is a "nest" of cells in the ovary within which the egg is formed. (4) An ovary is a female reproductive organ which produces eggs.

60. **(4)** Sperm are deposited in the vagina during coitus. INCORRECT CHOICES: (1) A represents a follicle in the ovary from which an egg has just been released by ovulation. (2) Sperm cells must swim through E (cervix), D (uterus), and C (oviduct) to reach the egg. (3) Sperm cells must swim through C and D after being deposited in the vagina.

61. **(1)** C represents fertilization, the fusion of an egg and sperm. INCORRECT CHOICES: (2) Cleavage refers to the mitotic divisions of the animal embryo which lead to the formation of a blastula. (3) Implantation refers to the process by which the embryo becomes attached to and embedded within

the uterine lining. (4) Menstruation is the periodic shedding of the uterine lining which occurs when fertilization does not take place.

62. **(2)** Ovulation is the process by which an egg is released from the ovary. INCORRECT CHOICES: (1) Differentiation refers to the formation of specialized tissues and organs from the embryonic layers. (3) Gastrulation refers to the formation of the gastrula, an embryo with three embryonic layers. (4) Germination refers to the growth of a seed into a young plant.

63. **(4)** The fetus becomes implanted in the uterine lining and continues development within the uterus. INCORRECT CHOICES: (1) *E* represents the cervix of the uterus where the uterus is connected to the vagina. (2) *F* represents the vagina or birth canal through which the baby will be born. (3) *C* represents the oviduct where fertilization and cleavage take place.

64. **(3)** Yeasts carry on a form of anaerobic respiration which produces carbon dioxide, ethyl alcohol, and 2 ATP. INCORRECT CHOICES: (1), (2), and (4) Hydrae, earthworms, and grasshoppers obtain energy through aerobic respiration of glucose.

65. **(2)** Fermentation is a form of anaerobic respiration which produces carbon dioxide, ethyl alcohol, and 2 ATP. INCORRECT CHOICES: (1) Aerobic respiration requires oxygen and yields 38 ATP. (3) In chemosynthesis, the energy from the chemical reaction is used in the formation of organic compounds. (4) Dehydration synthesis is the formation of large molecules from smaller ones by removal of water.

66. **(3)** Chitinous appendages is one of the characteristics of the phylum Arthropoda to which the grasshopper belongs. Such specialization then needs a coordinated set of muscles to initiate movement. INCORRECT CHOICES: (1) Hydra locomotes by use of its tentacles in a somersaulting motion or merely slides along on its basal disc. (2) The Paramecium belonging to the Ciliophora lacks any specialized structures for locomotion outside of its cilia. (4) The human moves by the interaction of bones and muscles.

67. **(2)** Nitrogenous wastes, in the form of ammonia and urea, are excreted by the action of paired nephridia located in each body segment of the earthworm. INCORRECT CHOICES: (1) Carbon dioxide is excreted by diffusion through the moist skin of the earthworm; glycerol is a raw material in the synthesis of many lipids. (3) Feces and bile salts are egested from the large intestine through the anus in humans. (4) Amino acids are the building blocks of proteins but when broken down yield ammonia and urea as waste products; molecular nitrogen is not a waste product of animals although many compounds of nitrogen are.

68. **(2)** Two amino acids are bonded by the peptide bond at 3. INCORRECT CHOICES: (1), (3), and (4) An amino acid has both an amino group ($-NH_2$) and a carboxyl group ($-COOH$) bonded to the same carbon atom; in this fragment, one amino acid has its amino group at 1 and its carboxyl group is bonded to the amino group of the second amino acid at 3, while the carboxyl group of the second amino acid is at 4.

69. **(3)** The C-N bond (peptide bond) at 3 is a bond between the amino group of one amino acid and the carboxyl group of another. INCORRECT CHOICES: (1) 1 represents an amino group. (2) 2 represents a methyl group. (4) 4 represents a carboxyl group.

70. **(2)** 2 represents a methyl group; this part of an amino acid may vary from molecule to molecule and, for example, in the residue of an amino acid on the right it is a single hydrogen atom. INCORRECT CHOICES: (1) Every amino acid contains an amino group. (3) 3 represents the peptide bond between two amino acids. (4) Every amino acid contains a carboxyl group.

Topic Analysis Key

If you had these answers wrong	Study these chapters:
7,13,29,45,46,47,48,68,69,70	The Study of Life
1,5,6,27,28,42,43,44,66,67	Maintenance in Animals
4,24,25,26,30,38,40,41,45,46,47, 64,65	Maintenance in Plants
3,21,22,23,37,59,60,61,62,63	Reproduction and Development
2,8,9,18,19,20,35,56,57,58	Genetics
10,11,17,31,32,33,34,36,54,55	Evolution and Diversity
12,14,15,16,39,49,50,51,52,53	Plants and Animals in Their Environment.

Biology Review

The following review material for Biology is based on the New York State Syllabus and features actual previous Regents questions. The chapter overviews are prepared in an outline format, in order to present the topics in clear terms: easy to follow and easy to learn.

Although the material does not go into depth within each topic area (that is the function of your textbook), you will be able to determine those areas that are essential for scoring high on your Biology Regents Examination.

After evaluating your strengths and weaknesses with the Pretest, you will be able to determine which areas need additional work. Read each overview carefully. Then practice the questions in the appropriate chapters that follow.

1

The Study of Life (Biochemistry)

A. The Concept of Life
  1. Definition of Life
  2. Activities of Life
    a. Nutrition
    b. Transport
    c. Respiration
    d. Excretion
    e. Synthesis
    f. Regulation
    g. Growth
    h. Reproduction

B. The Units of Life
  1. Cell theory
    a. Historical background
    b. Currently accepted concepts
    c. Exceptions to concepts
  2. Cell study
    a. Investigational techniques
      (1) Instruments used in cell study
      (2) Measurement
    b. Structural components

C. The Chemistry of Life
  1. Chemical elements in living matter

2. Chemical compounds in living matter
   a. Inorganic compounds
   b. Organic compounds
      (1) Carbohydrates
         (a) Composition
         (b) Structure
      (2) Lipids
         (a) Composition
         (b) Structure
      (3) Proteins
         (a) Composition
         (b) Structure
      (4) Nucleic Acids
         (a) Composition
         (b) Structure

3. Chemical activity in living matter
   a. The role of enzymes
      (1) Structure
         (a) Protein nature
         (b) Active site
      (2) Function
         (a) Enzyme-substrate complex
         (b) "Lock-and-key" model
      (3) Factors influencing enzyme action
         (a) pH
         (b) Temperature
         (c) Relative amounts of enzyme and substrate
   b. Reactions
      (1) Dehydration synthesis
      (2) Hydrolysis

The following questions on Biochemistry and the Study of Life have appeared on previous Regents examinations.

Directions: Each question is followed by four choices. Underline the correct choice.

Directions: Base your answer to question 1 on the structural formula below and on your knowledge of biology.

1. The structural formula represents a molecule of
   (1) glucose
   (2) glycerol
   (3) maltose
   (4) alanine

2. A rotten egg may give off a foul-smelling gas containing sulfur. Which decomposing chemical compounds in the egg are most likely the source of this odor?
   (1) proteins
   (2) nucleic acids
   (3) carbohydrates
   (4) lipids

3. Starch is classified as a
   (1) disaccharide
   (2) polypeptide
   (3) nucleotide
   (4) polysaccharide

4. What two molecules are produced when two glucose molecules are chemically bonded together?
   (1) a lipid and an enzyme
   (2) a polypeptide and oxygen
   (3) a polysaccharide and carbon dioxide
   (4) a disaccharide and water

5. Which process is represented below?

   simple organic molecules $\xrightarrow{\text{enzymes}}$ complex organic molecules + H_2O
   (1) hydrolysis
   (2) synthesis
   (3) digestion
   (4) respiration

6. Which chemical activity is correctly matched with the waste product it produces?
   (1) protein metabolism—urea
   (2) anaerobic respiration—oxygen
   (3) photosynthesis—carbon dioxide
   (4) dehydration synthesis—carbon dioxide

Directions: Base your answers to questions 7 and 8 on the formula below and on your knowledge of biology.

7. The diagram represents the structural formula of a specific
   (1) inorganic acid
   (2) fatty acid
   (3) nucleic acid
   (4) amino acid

8. The entire group within the circled area is known as
   (1) a carboxyl group
   (2) a methyl group
   (3) an amino group
   (4) a hydroxyl group

9. The excretory organelles of some unicellular organisms are contractile vacuoles and
   (1) cell membranes
   (2) cell walls
   (3) ribosomes
   (4) centrioles

10. Which cell structure contains respiratory enzymes?
    (1) cell wall
    (2) nucleolus
    (3) mitochondrion
    (4) vacuole

11. Which is the principal inorganic compound found in cytoplasm?
    (1) lipid
    (2) carbohydrate
    (3) water
    (4) nucleic acid

12. To separate the parts of a cell by differences in density, a biologist would probably use
    (1) a microdissection instrument
    (2) an ultracentrifuge
    (3) a phase-contrast microscope
    (4) an interference microscope

13. The letter "p" as it normally appears in print is placed on the stage of a compound light microscope. Which best represents the image observed when a student looks through the microscope?

    (1) p   (2) q   (3) b   (4) d

14. The cellular function of the endoplasmic reticulum is to
    (1) provide channels for the transport of materials
    (2) convert urea into a form usable by the cell
    (3) regulate all cell activities
    (4) change light energy into chemical bond energy

15. In which life function is the potential energy of organic compounds converted to a form of stored energy which can be used by the cell?
    (1) transport
    (2) respiration
    (3) excretion
    (4) regulation

16. The compound whose structural formula is shown below is a building block of what class of organic compounds?

    (1) proteins
    (2) carbohydrates
    (3) lipids
    (4) starches

17. Which occurs when two glucose molecules are chemically bonded and form a molecule of maltose?
    (1) hydrolysis of a protein
    (2) the release of a water molecule
    (3) the addition of a carbon molecule
    (4) the formation of a starch molecule

18. Which isotope has been used to investigate the photochemical reactions involved in photosynthesis?
    (1) phosphorus-32
    (2) sulfur-35
    (3) nitrogen-15
    (4) oxygen-18

19. The synthesis of ATP molecules from ADP molecules takes place chiefly in the
    (1) endoplasmic reticulum
    (2) ribosomes
    (3) mitochondria
    (4) Golgi bodies

20. An isotope is used to trace the chemical reactions of photosynthesis in a green plant. If the isotope is ultimately found in the starch stored in the green plant, the isotope would be
    (1) nitrogen-14
    (2) sulfur-32
    (3) carbon-14
    (4) phosphorus-35

21. Two of the most common storage products in animals are fat and
    (1) glucose
    (2) lipase
    (3) maltose
    (4) glycogen

22. The process by which proteins are made from amino acids is known as
    (1) dehydration synthesis
    (2) intracellular digestion
    (3) ingestion
    (4) hydrolysis

23. Which process is illustrated by the summary equation below?

    glucose + oxygen $\xrightarrow{\text{enzymes}}$ water + carbon dioxide + 38 ATP
    (1) hydrolysis
    (2) dehydration synthesis
    (3) photosynthesis
    (4) aerobic respiration

24. A nitrogenous waste product resulting from the metabolism of protein molecules is
    (1) carbon dioxide
    (2) ammonia
    (3) mineral salts
    (4) water

25. Which process would include a net movement of sugar molecules through a membrane from a region of lower concentration to a region of higher concentration?
    (1) osmosis
    (2) cyclosis
    (3) active transport
    (4) passive transport

26. Enzymes which speed up the hydrolysis of fats are known as
    (1) amylases
    (2) lipases
    (3) maltases
    (4) proteases

27. During a period of strenuous exercise, the muscle cells of humans may carry on anaerobic respiration and produce a high concentration of
    (1) glucose
    (2) carbon dioxide
    (3) lactic acid
    (4) ethyl alcohol

28. When a cell uses energy and moves materials across a cell membrane, the process is known as
    (1) osmosis
    (2) active transport
    (3) diffusion
    (4) passive transport

29. The complete digestion of animal starch results in the formation of
    (1) glucose molecules
    (2) amino acids
    (3) fatty acids
    (4) glycogen molecules

30. In the diagram of the cell below, which cell structure contains most of the enzymes necessary for cellular respiration.

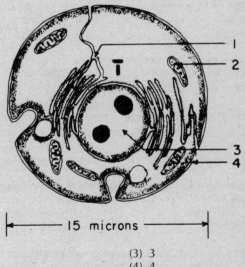

    (1) 1
    (2) 2
    (3) 3
    (4) 4

31. Which formula represents a carbohydrate?
    (1) $NaHCO_3$
    (2) $C_{18}H_{36}O_2$
    (3) NH_2CH_2COOH
    (4) $C_{12}H_{22}O_{11}$

32. Which element is present in all lipids and proteins?
    (1) iron
    (2) carbon
    (3) nitrogen
    (4) calcium

33. Which organelle contains hereditary factors and controls most cell activities?
    (1) nucleus
    (2) cell membrane
    (3) vacuole
    (4) endoplasmic reticulum

34. Organic molecules that are catalysts for chemical reactions occurring in living cells are
    (1) lipids
    (2) enzymes
    (3) carbohydrates
    (4) nucleic acids

35. Compared to the ingested food molecules, the endproduct molecules of digestion are usually
    (1) smaller and more soluble
    (2) larger and more soluble
    (3) smaller and less soluble
    (4) larger and less soluble

36. Which life activity is *not* required for the survival of an individual organism?
    (1) nutrition
    (2) respiration
    (3) reproduction
    (4) synthesis

37. A specific organic compound contains only the elements carbon, hydrogen, and oxygen in the ratio of 1:2:1. This compound is most probably a
    (1) nucleic acid
    (2) carbohydrate
    (3) protein
    (4) lipid

38. An important function of the ultracentrifuge is to
    (1) increase the magnification of different parts of cells
    (2) slice specimens embedded in wax into thin sections
    (3) transplant nuclei from one cell to another
    (4) separate different cell organelles by their density

39. The diagram below represents the field of vision of a microscope. What is the approximate diameter of the cell shown in the field?

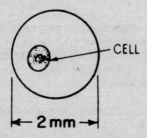

    (1) 50 microns
    (2) 500 microns
    (3) 1,000 microns
    (4) 2,000 microns

40. The formation of peptide bonds occurs as the result of
    (1) enzymatic hydrolysis
    (2) capillary action
    (3) translocation
    (4) dehydration synthesis

41. Glycogen is most similar in composition and structure to
    (1) potato starch
    (2) glycerol
    (3) hemoglobin
    (4) bacon fat

42. Centrioles are cell structures involved primarily in
    (1) cell division
    (2) storage of fats
    (3) enzyme production
    (4) cellular respiration

Directions: Base your answers to questions 43 and 44 on the graph below which represents data taken on the metabolic rate of a certain species of bacteria under a variety of temperature conditions.

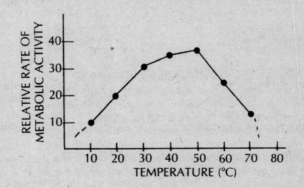

43. For this species of bacteria, temperatures above 70°C will probably
    (1) not affect life processes
    (2) increase the rate at which life processes occur
    (3) be optimum for life processes
    (4) stop life processes

44. Which temperature range is most advantageous for this species in terms of metabolic rate?
    (1) 5–15°C
    (2) 20–30°C
    (3) 40–50°C
    (4) 60–70°C

45. In an Ameba, which cell organelle is the site of protein hydrolysis?
    (1) ribosome
    (2) mitochondrion
    (3) cell wall
    (4) food vacuole

Directions: Base your answers to questions 46 and 47 on the formula shown below.

46. The portion of the formula that is enclosed by the dashed line is known as
    (1) a hydroxyl group
    (2) an amino group
    (3) a methyl group
    (4) a carboxyl group

47. The type of formula shown above is best classified as
    (1) a polymer formula
    (2) a structural formula
    (3) an inorganic formula
    (4) a metabolic formula

48. The diameter of a certain cell can be expressed as 0.007 millimeter or
    (1) 7 μ
    (2) 7 cm
    (3) 1000 μ
    (4) 0.07 inch

49. When a thin section of living plant tissue was mounted on a slide and observed under a compound light microscope, it appeared that some of the cells were without nuclei. Which is the most likely explanation for this observation?
    (1) Cell nuclei can be seen only with the phase contrast microscope.
    (2) All the nuclei were not at the same depth in the section.
    (3) The cells did not have enough light to form nuclei.
    (4) Some cells are too small to contain nuclei.

50. The most abundant inorganic compound in living systems is
    (1) water
    (2) mineral salts
    (3) carbon dioxide
    (4) ammonia

Directions (51-53): For each statement in questions 51 through 53, select the biochemical process, *chosen from the list below*, that is best described by that statement. [A number may be used more than once or not at all.]

Biochemical Processes

(1) Photosynthesis
(2) Replication
(3) Chemosynthesis
(4) Fermentation

51. Solar energy is converted to chemical energy in organic molecules.

52. Some bacteria obtain energy for food production by oxidizing compounds of sulfur or iron.

53. Lactic acid is produced as a result of this process.

54. Proteins are synthesized from less complex organic compounds known as
    (1) enzymes
    (2) starches
    (3) carbons
    (4) amino acids

55. Small, soluble food molecules are converted to larger, insoluble molecules by the process of
    (1) hydrolysis
    (2) respiration
    (3) synthesis
    (4) fermentation

56. Which formula represents an organic compound?
    (1) NH_3
    (2) H_2O
    (3) NaCl
    (4) $C_{12}H_{22}O_{11}$

Directions: Base your answers to questions 57 and 58 on the two compounds whose formulas are shown below.

57. Which element is present in both of the molecules shown, but *not* present in maltose?
    (1) carbon
    (2) nitrogen
    (3) hydrogen
    (4) oxygen

58. When the two molecules are joined together chemically, a molecule of water is released. This process is known as
    (1) dehydration synthesis
    (2) hydrolysis
    (3) absorption
    (4) transpiration pull

59. The ribosomes of plant cells are sites for the synthesis of
    (1) ATP
    (2) sugars
    (3) nucleic acids
    (4) enzymes

60. Which microscope magnification should be used to observe the largest field of view of an insect wing?
    (1) $20\times$
    (2) $100\times$
    (3) $400\times$
    (4) $900\times$

61. Centrosomes are normally present in the
    (1) cytoplasm of onion cells
    (2) cytoplasm of cheek cells
    (3) nuclei of liver cells
    (4) nuclei of bean cells

62. The reactions involving most chemical compounds in living systems depend upon the presence of
    (1) sulfur as an enzyme
    (2) water as a solvent
    (3) salt as a substrate
    (4) nitrogen as an energy carrier

63. Which substance is an inorganic compound that is necessary for most of the chemical reactions to take place in living cells?
    (1) glucose
    (2) starch
    (3) water
    (4) amino acid

64. If a specific carbohydrate molecule contains ten hydrogen atoms, that same molecule would most probably contain
    (1) one nitrogen atom
    (2) ten nitrogen atoms
    (3) five oxygen atoms
    (4) twenty oxygen atoms

65. Intracellular transport of materials is most closely associated with which cell organelle?
    (1) cell membrane
    (2) cell wall
    (3) ribosome
    (4) endoplasmic reticulum

66. According to the chart below, which neuron is the longest?

Neuron	Length of Neuron
A	1.5 microns
B	50.0 microns
C	0.5 millimeter
D	0.005 millimeter

(1) A   (3) C
(2) B   (4) D

67. Enzymes are produced as a direct result of which process?
(1) protein synthesis   (3) respiration
(2) photosynthesis   (4) enzymatic hydrolysis

68. Which arrangement of atoms represents a group which is characteristic of organic acids?

(1) H—C—H (with H above and below the C)

(2) —C with double bond O and OH

(3) —N with H and H

(4) —OH

69. The removal of metabolic wastes from living cells is known as
(1) secretion   (3) excretion
(2) ingestion   (4) digestion

70. Which term is used to represent all of the physiological activities carried on by an organism?
(1) regulation   (3) homeostasis
(2) metabolism   (4) synthesis

Directions: Base your answers to questions 71 through 74 on the graph below which represents the rate of action of four different enzymes at varying temperatures.

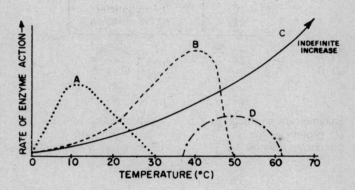

71. For which enzyme is the maximum rate of enzyme action achieved at 10°C?
    (1) *A*   (3) *C*
    (2) *B*   (4) *D*

72. Which enzyme begins its action at the highest temperature?
    (1) *A*   (3) *C*
    (2) *B*   (4) *D*

73. Which curve most closely illustrates the effect of temperature on human enzymes?
    (1) *A*   (3) *C*
    (2) *B*   (4) *D*

74. Which curve most likely represents incorrect data for the rate of enzyme action?
    (1) *A*   (3) *C*
    (2) *B*   (4) *D*

Answers

1. 1	20. 3	39. 2	58. 1
2. 1	21. 4	40. 4	59. 4
3. 4	22. 1	41. 1	60. 1
4. 4	23. 4	42. 1	61. 2
5. 2	24. 2	43. 4	62. 2
6. 1	25. 3	44. 3	63. 3
7. 4	26. 2	45. 4	64. 3
8. 1	27. 3	46. 4	65. 4
9. 1	28. 2	47. 2	66. 3
10. 3	29. 1	48. 1	67. 1
11. 3	30. 2	49. 2	68. 2
12. 2	31. 4	50. 1	69. 3
13. 4	32. 2	51. 1	70. 2
14. 1	33. 1	52. 3	71. 1
15. 2	34. 2	53. 4	72. 4
16. 1	35. 1	54. 4	73. 2
17. 2	36. 3	55. 3	74. 3
18. 4	37. 2	56. 4	
19. 3	38. 4	57. 2	

Explanatory Answers

1. **(1)** Glucose ($C_6H_{12}O_6$) has the structural formula shown here. INCORRECT CHOICES: (2) Glycerol is a 3-carbon alcohol with the formula, $C_3H_8O_5$. (3) Maltose is a disaccharide with the formula, $C_{12}H_{22}O_{11}$. (4) Alanine is an amino acid with the formula, $C_3H_7O_2N$.

2. **(1)** Proteins may contain sulfur and when they decompose may liberate or give off hydrogen sulfide which smells like rotten egg. INCORRECT CHOICES: (2) Nucleic acid molecules contain the elements carbon, oxygen, hydrogen, nitrogen, and phosphorus. (3) Carbohydrates contain a ratio of one carbon atom, to two hydrogen atoms, to one oxygen atom. (CH_2O). (4)

Lipid molecules contain a ratio of one carbon atom, to one hydrogen atom, to one oxygen atom (CHO).

3. **(4)** Starch is a chain of many simple sugars. INCORRECT CHOICES: (1) A disaccharide is composed of only two simple sugar molecules. (2) A polypeptide is a chain of amino acids. (3) A nucleotide is composed of a phosphate, a five-carbon sugar, and a nitrogen base.

4. **(4)** In the dehydration synthesis of two glucose molecules, a double sugar or disaccharide and water are formed. INCORRECT CHOICES: (1) A lipid is formed from three fatty acids and one glycerol molecule; and an enzyme is made from amino acids. (2) A polypeptide is a chain of amino acids. Oxygen is formed during photosythesis. (3) A polysaccharide is a chain of more than two glucose molecules. Carbon dioxide is formed during respiration.

5. **(2)** Synthesis is the formation of larger, complex molecules from smaller, simpler ones. INCORRECT CHOICES: (1) Hydrolysis is the decomposition of larger molecules into smaller units by chemical combination with water. (3) Digestion is the process of breaking down large, insoluble molecules into smaller soluble ones. (4) Respiration is the oxidation of food which results in the release of energy.

6. **(1)** Urea is a waste product of the digestion of protein molecules. INCORRECT CHOICES: (2) Carbon dioxide is one waste product of anaerobic respiration. (3) Oxygen is produced during photosynthesis. (4) Water is produced from the process of dehydration synthesis.

7. **(4)** An amino acid is an organic compound containing an amino group and a carboxyl or acid group. INCORRECT CHOICES: (1) An inorganic acid would not contain both carbon and hydrogen, nor an amino group. (2) A fatty acid does not have the amino group. (3) A nucleic acid contains a phosphate group, a nitrogen base, and a five-carbon sugar.

8. **(1)** A carboxyl group or an organic acid group is depicted in the circle. INCORRECT CHOICES: (2) A methyl group contains a carbon atom surrounded by three hydrogen atoms. (3) An amino group is a nitrogen atom surrounded by two hydrogen atoms. (4) A hydroxyl group is a radical composed of one oxygen atom bound to one hydrogen atom.

9. **(1)** The cell membrane is an organelle through which waste gases and liquids can pass by active or passive transport. INCORRECT CHOICES: (2) Cell walls are nonliving structures which provide protection and give shape to the cells. (3) Ribosomes are organelles where proteins are synthesized.

(4) Centrioles are rodlike structures in the centrosome which have a role in cell division.

10. **(3)** The mitochondrion is the organelle containing respiratory enzymes which regulate the release of energy in the cell. INCORRECT CHOICES: (1) The cell wall is the nonliving structure surrounding plant cells. (2) The nucleolus is located in the nucleus of a cell. (4) A vacuole is a sac in a cell in which liquids are stored.

11. **(3)** The inorganic compound, water, comprises more than 85% of the cytoplasm. INCORRECT CHOICES: (1) A lipid is an organic molecule which is a fat, oil or wax. (2) A carbohydrate is a term used to describe an organic molecule of starch, sugar, or fat. (4) RNA and DNA are organic compounds called nucleic acids.

12. **(2)** After being centrifuged, the densest material is located at the bottom of the centrifuge tube; less dense materials will form layers or stratify above the denser materials. Cell parts of varying density can be separated in this manner using an ultracentrifuge. INCORRECT CHOICES: (1) To be cut apart, small objects are separated or dissected with the use of a microdissection instrument. (3) The images of objects in a phase-contrast microscope appear more clearly due to the angle of light passing through the specimen. (4) The image seen in an interference microscope varies in light intensity with reflection or refraction of light from the object.

13. **(4)** The image observed in the field of the microscope is reversed and inverted due to the action of the lenses used. INCORRECT CHOICES: (1) This image is upright and not reversed. (2) This image is reversed but not inverted. (3) This image is inverted but not reversed.

14. **(1)** The endoplasmic reticulum provides the pathways for the transport of materials in the cells. INCORRECT CHOICES: (2) Urea is a waste product from the breakdown of proteins. (3) The nucleus usually regulates all the cell activities. (4) Chlorophyll in green plants changes light energy into chemical bond energy.

15. **(2)** Respiration converts potential energy to stored energy of ATP; this form is readily available for the cell. INCORRECT CHOICES: (1) Transport is the taking in and distribution of materials within an organism. (3) Excretion is the removal of the waste products of metabolism. (4) Regulation is the control of all the activities of the organism.

16. **(1)** In looking at the diagram, we note the appearance of the amino group which is a clue to the appearance of amino acids; we see the appearance

of the carboxyl group which again points to amino acids. In addition, the elements carbon, hydrogen, oxygen, and nitrogen appear which is another clue to the appearance of amino acids. Amino acids of course build proteins. INCORRECT CHOICES: (2) This could not be a carbohydrate because of the appearance of the nitrogen element. (3) Lipids are composed of fatty acids, and the glycerol molecule, neither of which appear in the diagram. (4) Starches are polymers of glucose units, and again, glucose is a carbohydrate.

17. **(2)** The release of the water molecule is typical of the process of dehydration which is the process of synthesis to which the question refers. INCORRECT CHOICES: (1) Hydrolysis would be the breakdown of maltose into the two glucose units by the addition of water. (3) The addition of a carbon molecule would change the chemical and physical properties of the substance. Therefore we wouldn't be talking about glucose or maltose. (4) The formation of starch would involve the dehydration of many units of glucose.

18. **(4)** Oxygen-18 is substituted for normal oxygen in the water molecule to determine the path that oxygen takes in the reaction. It was found that the isotope appeared as liberated oxygen. When O^{18} was substituted in the CO_2, it was found that the heavy oxygen appeared in the sugar molecule. INCORRECT CHOICES: (1),(2),and(3) None of these elements appear in any of the reactants or products of the photosynthetic process and therefore could not be correct.

19. **(3)** The mitochondria are the sites of cellular respiration. Energy is made available to the cell in the form of ATP which is synthesized from ADP molecules in the mitochondria. INCORRECT CHOICES: (1) The endoplasmic reticulum is a transport organelle which forms meandering canals through the cytoplasm. (2) The ribosomes are protein factories. (4) The Golgi bodies are thought to synthesize carbohydrates and package secretions.

20. **(3)** Starch molecules are composed of polymers of glucose units. Since carbon dioxide is a reactant in photosynthesis and carbon is an element in starch, it would be conceivable that the carbon-14 isotope would be found in starch molecules. INCORRECT CHOICES: (1) Nitrogen is a component of amino acids and ultimately proteins. There are no nitrogen molecules in starch. (2) Sulfur is also a component of some amino acids. There is no sulfur in starch. (4) Phosphorus as an element is metabolized by the body in several reactions. There is no phosphorus in starch.

21. **(4)** Glycogen is a polysaccharide and is animal fat. INCORRECT CHOICES: (1)Glucose is not a storage product but the simplest carbohydrate. (2)

Lipase is an enzyme which catalyzes the hydrolysis of lipids. (3) Maltose is a dissacharide and not a storage product.

22. **(1)** The process of synthesis of any complex substance is always by dehydration. INCORRECT CHOICES: (2) Intracellular digestion involves the enzymatic hydrolysis of food within cellular organelles such as vacuoles. (3) Ingestion is the taking in of food by an organism. (4) Hydrolysis is the opposite of the synthesis process stated in the question.

23. **(4)** The equation shows glucose combining with oxygen (oxidation) and a release of 38 molecules of ATP. Typically this is aerobic respiration. INCORRECT CHOICES: (1) Hydrolysis is the addition of water to a substance to facilitate its breakdown. Water does not appear on the reactant side of the equation. (2) Dehydration is the removal of water in the presence of enzymes to synthesize a more complex substance. The reaction shows the hydrolysis of glucose not the synthesis involving glucose. (3) The photosynthetic reaction requires the presence of carbon dioxide, water, sunlight, and chlorophyll. These do not appear in this reaction.

24. **(2)** Ammonia is the result of deamination of proteins. The amino group is broken down into NH_3 and a free hydrogen atom. INCORRECT CHOICES: (1) Carbon dioxide is a waste product of respiration. (3) Mineral salts are formed by general body metabolism. (4) Water is a product of dehydration synthesis and is generally not considered to be a waste product.

25. **(3)** Active transport is a term which implies the use of energy to accomplish the transport of the substance. The situation as explained in the question is a movement of material against the concentration gradient and therefore requires the expenditure of energy. INCORRECT CHOICES: (1) Osmosis is the passive transport of water. (2) Cyclosis is the streaming motion of the cytoplasm. (4) Passive transport is the movement of molecules along the concentration gradient from areas of high concentration to low concentration.

26. **(2)** Lipases are enzymes which digest fats. Breaking the word down into its components we find the enzyme suffix, -ase, and the prefix lip- which begins the term lipid (fat). INCORRECT CHOICES: (1) Amylases digest starch compounds. (3) Maltases aid in the hydrolyzing of maltose to two glucose units. (4) Proteases aid in the hydrolyzation of proteins into amino acid molecules.

27. **(3)** During strenuous exercise, oxygen is being depleted faster in the muscle cells than the lungs can deliver. Therefore, incomplete respiration or oxidation of glucose occurs resulting in the accumulation of lactic acid.

INCORRECT CHOICES (1) Glucose is the reactant in both aerobic and anaerobic respiration. (2) Carbon dioxide is released in the Krebb's cycle which is part of the aerobic respiration process. (4) Ethyl alcohol is a product of anaerobic respiration by yeast cells.

28. **(2)** The use of ATP to facilitate absorption characteristically shows reverse concentration. Material must move against the concentration gradient and therefore energy is applied to accomplish this task. INCORRECT CHOICES: (1) Osmosis is the diffusion of water. (3) Diffusion is passive transport of dissolved materials. (4) Passive transport requires no energy expenditure by the cell.

29. **(1)** Glucose molecules represent the simplest end products of dehydration synthesis of starch. INCORRECT CHOICES: (2) Amino acids are the end products of protein metabolism. (3) Fatty acids are one end product of the metabolism of a lipid molecule. (4) Glycogen molecules represent animal fat.

30. **(2)** The diagram indicates the mitochondria where energy molecules are synthesized. The mitochondria is recognized by the double-folded membrane called the cristae. INCORRECT CHOICES: (1) The endoplasmic reticulum is recognized by its canal-like structure. (3) The nucleus is recognized by being somewhat centralized in the cytoplasm and having a membrane around its surface. (4) The cell membrane is recognized as the interface between the external environment and the cytoplasm.

31. **(4)** A carbohydrate is organic, and contains the elements carbon, hydrogen, and oxygen. The hydrogen ratio to oxygen atoms is 2:1. INCORRECT CHOICES: (1) A carbohydrate does not possess the element sodium as does the example. (2) The compound has a hydrogen to oxygen ratio greater than 2:1 and therefore could not be a carbohydrate. (3) This compound contains the amino group (NH_2) and a carboxyl group (COOH) which are not found in carbohydrates.

32. **(2)** Carbon is the element of "life." It completes the bridge between the living and the nonliving. Since both lipids and proteins are organic structures and carbon is present in all organics, this choice seems simple. INCORRECT CHOICES: (1) Iron is an inorganic element. It is not present in either lipids or proteins. (3) Nitrogen is present in proteins but not lipids. (4) Calcium is present in neither lipids nor proteins.

33. **(1)** The nucleus is the brain of the cell, controlling all cellular activities of the cell and it contains the hereditary material in the chromosomes. INCORRECT CHOICES: (2) The cell membrane forms the interface between

the external environment and the internal environment of the cell. (3) Vacuoles perform many functions, two of which are digestion and excretion. (4) The endoplasmic reticulum forms a transport mechanism throughout the cell, extending from the nuclear membrane to the cell membrane.

34. **(2)** Enzymes are organic (protein) catalysts for reactions in living cells. INCORRECT CHOICES: (1) Lipids include fats, oils, and waxes none of which act as catalysts. (3) Carbohydrates include sugars, starch, and cellulose which do not act as catalysts. (4) Nucleic acids include DNA and RNA; DNA is the genetic material in cells.

35. **(1)** To digest means to break into more soluble and smaller units. If ingested food were not digested it would not be able to be assimilated by the organism. INCORRECT CHOICES: (2) Larger molecules could not readily pass through the membranes surrounding food vacuoles or through membranes of stomachs or intestines. (3) Digestion does provide for smaller molecules but it also provides for more solubility in that the passage of the molecules would be hindered if less soluble conditions prevailed. (4) Larger molecules than those of the ingested particles would mean synthesis has occurred and digestion is the opposite of synthesis. The meaning of less soluble has already been stated.

36. **(3)** Reproduction is required for the survival of the species but not the individual organism. INCORRECT CHOICES: (1) Without nutrition, the organism would die of starvation in a short time. (2) Respiration, the exchange of gases, is vital if the organism is going to obtain energy to carry out the life activities. (4) Synthesis is necessary to organize the simple substances derived from ingestion into more complex substances that the organism can use.

37. **(2)** Carbohydrates are distinguished from other organics by their component elements and the ratio of those elements. INCORRECT CHOICES: (1) A nucleic acid is composed of nucleotide bases which in turn are composed of phosphate groups, sugars, and nitrogenous bases. (3) A protein has the following elements: carbon, hydrogen, oxygen, nitrogen, and sometimes sulfur. (4) A lipid contains the same elements as the carbohydrate but the ratio is greater than the 2:1.

38. **(4)** An ultracentrifuge separates components according to their density. This aids cytologists in studying organelles which may otherwise not be detachable. INCORRECT CHOICES: (1) This choice refers to various microscope techniques. (2) The microtome slices tissues and cells. (3) Micro-

dissection equipment is employed to transplant delicate organelles such as the nucleus.

39. **(2)** The field of vision is 2000 microns in width. The cell occupies approximately one fourth of the width. Therefore, in microns, the cell would occupy about 500 microns of the width. INCORRECT CHOICES: (1) 50 microns is 1/40 of 2000. (3) 1000 is one half the width of the field. (4) 2000 is the total width of the field.

40. **(4)** Dehydration synthesis is the process by which large organic molecules are built up from their building blocks with the release of water molecules; peptide bonds are formed between amino acids during the dehydration synthesis of amino acids. INCORRECT CHOICES: (1) Enzymatic hydrolysis is the process by which large organic molecules are broken down into their building blocks. (2) Capillary action refers to the tendency of water to rise in very narrow tubes. (3) Translocation refers to circulation in plants.

41. **(1)** Glycogen and potato starch are both polysaccharides formed of many glucose molecules bonded together. INCORRECT CHOICES: (2) Glycerol is used in the synthesis of many lipids. (3) Hemoglobin is a protein synthesized from amino acids. (4) Bacon fat is a lipid synthesized from fatty acids and glycerol.

42. **(1)** Centrioles, found most often in animal cells, are involved with cellular reproduction. They contain the astral rays and spindle fibers. INCORRECT CHOICES: (2) Storage of fats in unicellular organisms may be a function of vacuoles. (3) Enzyme production is the responsibility of the ribosomes. (4) Cellular respiration occurs in the mitochondria.

43. **(4)** At temperatures above 70°C, the metabolic activity of the bacteria is zero, that is, life processes stop. INCORRECT CHOICES: (1) Metabolic activity includes all the life processes. (2) The rate at which life processes occur increases up to 50°C on the graph. (3) The highest relative rate of metabolic activity occurs at 50°C, so temperatures above 70°C would not be optimum for life processes.

44. **(3)** 40–50°C is the temperature range in which the relative rate of metabolic activity for this species is highest. INCORRECT CHOICES: (1) In the range 5–15°C, the rate of metabolic activity is low but increasing. (2) In the range 20–30°C, the rate of metabolic activity is increasing rapidly but has not reached its highest level. (4) In the range 60–70°C, the rate of metabolic activity is decreasing rapidly.

45. **(4)** Ingested food particles in the Ameba are surrounded by food vacuoles within which digestion (hydrolysis) of proteins occurs. INCORRECT CHOICES: (1) The ribosome is the site of protein synthesis. (2) The mitochondrion is the site of aerobic cellular respiration. (3) An Ameba has no cell wall which in a plant cell provides support and protection.

46. **(4)** The carboxyl (acid) group can also be written as $-COOH$. INCORRECT CHOICES: (1) A hydroxyl group is written $-OH$. (2) An amino group is written $-NH_2$. (3) A methyl group is written $-CH_3$.

47. **(2)** A structural formula is one which shows the position of the atoms making up an organic molecule. INCORRECT CHOICES: (1) A polymer is a molecule made up of many repeating units of one type. (3) The formula for an inorganic compound would contain no carbon. (4) Metabolism is the sum of all the chemical activities in an organism; the term, metabolic formula has no meaning.

48. **(1)** One millimeter equals 1000 microns, therefore, 0.007 millimeter equals 7 microns (μ). INCORRECT CHOICES: (2) 7 cm = 70 mm. (3) 1000μ = 1 mm. (4) 0.07 inch = 1.8 mm.

49. **(2)** Because the section of tissue was very thin, some nuclei which were located above or below the section were not included. INCORRECT CHOICES: (1) A phase contrast microscope is not needed to visualize nuclei. (3) Light is not needed for the formation of nuclei. (4) Even very small cells may contain nuclei.

50. **(1)** Water is an inorganic compound making up over two thirds of all living systems. INCORRECT CHOICES: (2) Mineral salts are found in small quantities in living systems. (3) Carbon dioxide is a waste product of cellular respiration and is excreted as its concentration rises. (4) Ammonia is a waste product of protein metabolism which is highly toxic and can be tolerated only at very low concentrations.

51. **(1)** The conversion of solar (radiant) energy to chemical energy is the express concern of photosynthetic organisms. INCORRECT CHOICES: (2) Replication is the duplication of chromosomes in mitosis. (3) Chemosynthesis is the food production of bacteria not utilizing the sun. (4) Fermentation is anaerobic respiration.

52. **(3)** In the absence of sunlight, some bacteria have adapted to synthesizing food from those elements listed. INCORRECT CHOICES: Refer to number 51.

53. **(4)** Fermentation produces lactic acid in muscle cells due to the incomplete breakdown of glucose. INCORRECT CHOICES: Refer to number 51.

54. **(4)** The building blocks of proteins are the amino acids. INCORRECT CHOICES: (1) Enzymes are organic catalysts which are protein in nature. (2) Starches are large carbohydrate molecules. (3) Carbons are many carbon atoms.

55. **(3)** Synthesis is the process of forming complex molecules from simple ones. INCORRECT CHOICES: (1) Hydrolysis is the chemical reaction in which large, insoluble molecules are broken down into small, soluble molecules. (2) Respiration is the process in which energy is released from organic compounds. (4) Fermentation is a form of anaerobic respiration.

56. **(4)** $C_{12}H_{22}O_{11}$ is an organic compound because it contains both carbon and hydrogen. INCORRECT CHOICES: (1) NH_3, ammonia, does not contain the element C. (2) H_2O or water also does not contain C. (3) NaCl, sodium chloride, is an inorganic salt.

57. **(2)** Both molecules are amino acids and contain nitrogen. Maltose is a sugar and does not contain nitrogen. INCORRECT CHOICES: (1), (3), and (4) Carbon, hydrogen, and oxygen are present in all sugars.

58. **(1)** Dehydration synthesis is the process of building large molecules from smaller ones by splitting out water. INCORRECT CHOICES: (2) Hydrolysis is breaking down large molecules by the addition of water. (3) Absorption is taking materials into a cell or organism. (4) Transpiration pull is the movement of water in a plant due to its loss through the leaves.

59. **(4)** Enzymes are always proteins; proteins are synthesized at the ribosomes. INCORRECT CHOICES: (1) ATP is a molecule which stores energy. (2) and (3) Sugars and nucleic acids are organic molecules and not organelles, which are parts of cells.

60. **(1)** Under the lowest magnification, the largest field can be seen; 20X is the lowest magnification of those given. INCORRECT CHOICES: (2) 100X provides 5X greater magnification than 20X, but shows only one fifth of the field. (3) Compared to 20X, only one half of the field can be seen under 40X. (4) Of the four choices, 900X gives the greatest magnification and the smallest field of view.

61. **(2)** Centrosomes are found only in the cytoplasm of animal cells. INCORRECT CHOICES: (1) Onion cells are plant cells; their cytoplasm does not

contain centrosomes. (3) Centrosomes are not found in the nucleus. (4) Centrosomes are never found in the nucleus or in plant cells.

62. **(2)** In order to react, most chemical compounds in the cell must be dissolved in water. INCORRECT CHOICES: (1) Enzymes are organic catalysts; sulfur is an element. (3) A substrate is a compound acted upon by an enzyme; salt is not a substrate. (4) Nitrogen is an element; it is not an energy carrier.

63. **(3)** Water (H_2O) is an inorganic compound making up a high percentage of all living matter. INCORRECT CHOICES: (1) and (2) Glucose and starch are organic compounds, carbohydrates. (4) Amino acids are organic compounds which serve as the building blocks of proteins.

64. **(3)** Hydrogen and oxygen are present in a 2:1 ratio in most carbohydrates; therefore, if 10 hydrogen atoms are present, there will be 5 oxygen atoms. INCORRECT CHOICES: (1) and (2) Carbohydrates do not contain nitrogen atoms. (4) A carbohydrate molecule with 20 oxygen atoms would have 40 hydrogen atoms.

65. **(4)** The endoplasmic reticulum provides channels through which transport of materials occurs in the cytoplasm. INCORRECT CHOICES: (1) The cell membrane controls the transport of materials into and out of the cell. (2) The cell wall supports and protects the plant cell. (3) The ribosome functions as the site for protein synthesis.

66. **(3)** One millimeter equals 1,000 microns. Neuron C is 0.5 millimeter long or 500 microns long. INCORRECT CHOICES: (1) Neuron A is 1.5 microns long. (2) Neuron B is 50.0 microns long. (4) Neuron D is 0.005 millimeter long or 5 microns long.

67. **(1)** All enzymes are either exclusively proteins or are proteins with attached, nonprotein, side groups. The protein portion of an enzyme is synthesized by the cell from amino acids. INCORRECT CHOICES: (2) Photosynthesis converts light energy into the chemical bond energy of carbohydrates. (3) Respiration converts the potential energy of organic molecules such as glucose to the more available form of ATP. (4) Enzymatic hydrolysis involves the breakdown of large organic molecules into their building blocks. This breakdown is under the control of enzymes.

68. **(2)** The carboxyl group ($-COOH$) is characteristic of organic acids. INCORRECT CHOICES: (1) $-CH_3$ represents a methyl group. (3) $-NH_2$ is an amino group found in amino acids. (4) $-OH$ represents an hydroxyl group.

69. **(3)** Excretion removes harmful, or potentially harmful, substances which form as waste products of metabolism in an organism. INCORRECT CHOICES: (1) Secretion involves the synthesis of substances with special metabolic functions and the release of these substances from the cell, tissue, or organ which produced them. (2) Ingestion involves the intake by an animal of those compounds and/or organisms which serve as nutrients. (4) Digestion is the process by which large, insoluble food particles are reduced to small, soluble molecules.

70. **(2)** Metabolism describes the sum of all the chemical activities that occur in an organism. INCORRECT CHOICES: (1) Regulation involves control of the various physiological activities of an organism. (3) Homeostasis refers to the ability of an organism to maintain a constant internal environment. (4) Synthesis involves those chemical activities by which complex molecules are created from simple compounds.

71. **(1)** Enzyme A reaches its highest point on the vertical axis of rate of enzyme action at 10°C. INCORRECT CHOICES: (2) Enzyme B reaches its highest point at 40°C. (3) Enzyme C shows an indefinite increase in rate of action. (4) Enzyme D reaches its highest rate at 50°C.

72. **(4)** Enzyme D begins action at about 37°C. INCORRECT CHOICES: (1), (2), and (3) Enzymes A, B, and C all begin activity at 0°C.

73. **(2)** Enzyme B reaches its highest rate of action at 40°C; average normal human body temperature is 37°C. INCORRECT CHOICES: (1) Enzyme A ceases activity above 30°C and so would not be active in the human body. (3) and (4) Enzymes C and D are not as active at 37°C, human body temperature, as is enzyme B.

74. **(3)** Proteins, including enzymes, are denatured at high temperatures. INCORRECT CHOICES: (1), (2), and (4) As temperature increases, the rate of enzyme action increases up to the relatively high temperature at which the shape of the enzyme molecule is altered by the heat. Above this temperature, the rate of enzyme action drops.

2

Maintenance in Animals
(Human Physiology)

A. Nutrition
   1. Process
      a. Ingestion
      b. Digestion
         (1) Mechanical aspects
         (2) Chemical aspects
            (a) Enzyme action
            (b) Outcome
   2. Adaptations
      a. Protozoa
      b. Hydra
      c. Earthworm
      d. Grasshopper
      e. Man
   3. Human Digestion
      a. Digestive system
         (1) The alimentary canal
         (2) Digestive glands

B. Transport
   1. Process
      a. Absorption
         (1) Structure of the cell membrane
         (2) Function of the cell membrane
            (a) Passive transport
            (b) Active transport and pinocytosis
   2. Adaptations
      a. Protozoa
      b. Hydra

      c. Earthworm
      d. Grasshopper
      e. Man
   3. Human circulation
      a. Circulatory system
         (1) Structures
            (a) Heart
            (b) Arteries
            (c) Capillaries
            (d) Veins
            (e) Lymph vessels
         (2) Fluids
            (a) Blood
            (b) Lymph
      b. Circulatory function
         (1) Transport
         (2) Protection
            (a) Clotting
            (b) Phagocytosis
            (c) Immunological reactions

C. Respiration
   1. Process
      a. Cellular level
         (1) Aerobic respiration
            (a) Description
            (b) Chemical aspects
         (2) Anaerobic respiration
            (a) Description
            (b) Chemical aspects
      b. Organism level
   2. Adaptations
      a. Protozoa
      b. Hydra
      c. Earthworm
      d. Grasshopper
      e. Man
   3. Human respiration
      a. Respiratory system
      b. Respiratory function

D. Excretion
   1. Process
   2. Adaptations

      a. Protozoa
      b. Hydra
      c. Earthworm
      d. Grasshopper
      e. Man
   3. Human excretion
      a. Excretory system
      b. Excretory function
         (1) Lungs
         (2) Liver
         (3) Skin
         (4) Kidney

E. Synthesis
   1. Process
   2. Products
      a. Secretions
         (1) Enzymes
         (2) Hormones
         (3) Neurohumors
         (4) Other
      b. Structural compounds
   3. Outcome
   4. Human synthesis
      a. Limitations
      b. Synthesis in relation to diet

F. Regulation
   1. Process
      a. Nervous control
         (1) Stimulus and response
         (2) Transmission of nerve impulses
            (a) Neurons
            (b) Synapses
            (c) Neurohumors
      b. Endocrine control
         (1) Hormones
         (2) Effects
      c. Comparison of nervous and endocrine systems
   2. Adaptations
      a. Nervous systems
         (1) Protozoa
         (2) Hydra

          (3) Earthworm
          (4) Grasshopper
          (5) Man
      b. Endocrine systems
  3. Human nervous regulation
      a. Nervous system
          (1) Structural units
              (a) Sensory neurons
              (b) Associative neurons
              (c) Motor neurons
          (2) Organization
              (a) Nerves
              (b) Ganglia and plexuses
              (c) Spinal cord
              (d) Brain: Cerebrum, Cerebellum, Medulla
      b. Nervous functions
          (1) Involuntary behavior
              (a) Reflexes
              (b) Conditioned behavior
  4. Human endocrine regulation
      a. Endocrine system
      b. Endocrine function
          (1) Thyroid gland
          (2) Parathyroid glands
          (3) Adrenal glands
          (4) Pituitary gland
          (5) Pancreas
          (6) Sex glands

G. Locomotion
  1. Process
  2. Adaptations
      a. Protozoa
      b. Hydra
      c. Earthworm
      d. Grasshopper
      e. Man
  3. Human locomotion
      a. Locomotion system
          (1) Bones
          (2) Muscles
          (3) Tendons
          (4) Ligaments
      b. Locomotive function

The following questions on Maintenance in Animals have appeared on previous Regents Examinations.

Directions: Each question is followed by four choices. Underline the correct choice.

1. Proteins not ingested by humans are found in human cells. The presence of these proteins is most directly a result of
   (1) regulation
   (2) synthesis
   (3) respiration
   (4) excretion

2. In the human organism, which part of the blood functions as a phagocyte?
   (1) red blood cells
   (2) platelets
   (3) white blood cells
   (4) pathogens

3. Oxygen carried by the blood in the capillaries normally enters body cells by
   (1) active transport
   (2) osmosis
   (3) diffusion
   (4) pinocytosis

4. Which process is represented by the summary equation below?

$$\text{glucose} \xrightarrow{\text{enzymes}} 2 \text{ lactic acid} + 2 \text{ ATP}$$

   (1) hydrolysis
   (2) dehydration synthesis
   (3) anaerobic respiration
   (4) aerobic respiration

5. Two end-products of digestion are
   (1) bile and enzymes
   (2) proteins and lipids
   (3) amino acids and simple sugars
   (4) proteins and carbohydrates

6. The cell membrane is composed mainly of
   (1) proteins and lipids
   (2) sugars and starches
   (3) cellulose
   (4) polysaccharides

7. Which are excretory organs?
   (1) skin and heart
   (2) lungs and kidneys
   (3) liver and testes
   (4) kidneys and pancreas

8. Endocrine glands produce chemical messengers known as
   - (1) enzymes
   - (2) neurohumors
   - (3) auxins
   - (4) hormones

9. When diffusion of certain substances through a cell membrane is *not* possible, the transport of these materials into the cell may be accomplished by
   - (1) passive transport
   - (2) osmosis
   - (3) cytoplasmic streaming
   - (4) pinocytosis

10. What is the function of oxygen in aerobic respiration?
   - (1) It forms pyruvic acid.
   - (2) It forms glucose.
   - (3) It acts as the final acceptor of hydrogen.
   - (4) It changes proteins to amino acids.

Directions: Base your answers to questions 11 through 13 on the diagram below of a nephron and its capillaries and on your knowledge of biology.

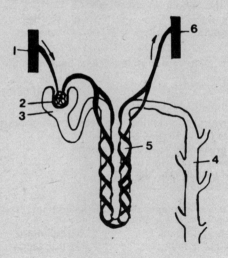

11. Into which structure does the filtrate first pass?
   - (1) 5
   - (2) 6
   - (3) 3
   - (4) 4

12. In which area is water being reabsorbed?
    (1) 5   (3) 3
    (2) 2   (4) 4

13. In which area does urine collect?
    (1) 1   (3) 6
    (2) 2   (4) 4

14. Glands which are part of the human digestive system, but *not* part of the alimentary canal are the
    (1) liver and stomach   (3) pancreas and small intestine
    (2) liver and pancreas   (4) pancreas and stomach

15. In which organism are materials transported throughout the body by an open circulatory system?
    (1) Hydra   (3) grasshopper
    (2) earthworm   (4) human

16. Which activity of life includes the absorption and distribution of essential materials throughout an organism?
    (1) excretion   (3) synthesis
    (2) locomotion   (4) transport

17. In humans, voluntary muscle tissue is located in the
    (1) brain   (3) legs
    (2) heart   (4) small intestine

18. The earthworm is classified as a heterotroph because it
    (1) ingests inorganic raw materials
    (2) uses preformed organic compounds
    (3) synthesizes proteins
    (4) fixes nitrogen from the soil

19. An animal has increased opportunities to obtain food as a direct result of
    (1) circulation   (3) synthesis
    (2) locomotion   (4) excretion

20. Ganglia are composed of clusters of
    (1) smooth muscle cells   (3) phagocytes
    (2) neurons   (4) striated muscle cells

21. Chemicals produced by the ends of neurons and secreted into synaptic spaces are
    (1) hormones
    (2) auxins
    (3) neurohumors
    (4) toxins

22. Transport into and out of the cells of a Hydra occurs by means of
    (1) osmosis, diffusion, and active transport
    (2) transpiration, cyclosis, and pinocytosis
    (3) a closed circulatory system
    (4) an open circulatory system

23. In which organ does peristalsis occur?
    (1) liver
    (2) pancreas
    (3) oral cavity
    (4) esophagus

24. The major function of a motor neuron is to
    (1) transmit impulses from the spinal cord to the brain
    (2) act as a receptor for environmental stimuli
    (3) transmit impulses from sense organs to the central nervous system
    (4) transmit impulses from the central nervous system to muscles or glands

25. The end products of digestion enter the fluids of animals through membranes by the process of
    (1) osmosis
    (2) absorption
    (3) synthesis
    (4) photosynthesis

26. Locomotion in the earthworm is accomplished through the action of setae and
    (1) flagella
    (2) cilia
    (3) bones
    (4) muscles

27. Which organism has a nervous system consisting of a simple anterior "brain" and a ventral nerve cord?
    (1) an earthworm
    (2) a Hydra
    (3) an Ameba
    (4) a human

28. A substance that is secreted by a gland and can regulate metabolism is a
    (1) nucleic acid
    (2) nitrogenous base
    (3) hormone
    (4) neurohumor

29. Which organism produces uric acid as its principal nitrogenous waste product?
    (1) Paramecium   (3) Ameba
    (2) Hydra   (4) grasshopper

30. In which organisms are hydrolytic enzymes present?
    (1) nongreen autotrophs, only
    (2) green autotrophs, only
    (3) heterotrophs, only
    (4) both heterotrophs and autotrophs

31. Secretions from ductless glands are known as
    (1) enzymes   (3) lachrymal fluids
    (2) hormones   (4) excretory fluids

32. Which organism *lacks* a one-way (tube within a tube) digestive system?
    (1) Hydra   (3) grasshopper
    (2) earthworm   (4) human

33. Changes in the environment that bring about responses are known as
    (1) effectors   (3) impulses
    (2) receptors   (4) stimuli

34. A similarity of the human nervous and endocrine systems is that both normally
    (1) secrete chemical messengers
    (2) have the same rate of response
    (3) have the same duration of response
    (4) secrete hormones that travel by way of neurons

35. A Paramecium, a Hydra, and an earthworm are similar in that they all
    (1) are sessile
    (2) are photosynthetic
    (3) have a closed transport system
    (4) have an external respiratory surface

36. Animals can *not* synthesize nutrients from inorganic raw materials. Therefore, animals obtain their nutrients by
    (1) combining carbon dioxide with water
    (2) consuming preformed organic compounds
    (3) hydrolyzing large quantitites of simple sugars
    (4) oxidizing inorganic molecules for energy

Directions 37–39: For each of questions 37 through 39, select the excretory structure, *chosen from the list below,* that best answers the question. Then record its *number* in the space next to the choice.

Excretory Structures
(1) Alveolus
(2) Nephron
(3) Sweat gland
(4) Liver

37. Which structure forms urine from water, urea, and salts?

38. Which structure removes carbon dioxide and water from the blood?

39. Which structure is involved in the breakdown of red blood cells?

Directions: Base your answers to questions 40 through 44 on the diagram of the adult human heart and on your knowledge of biology.

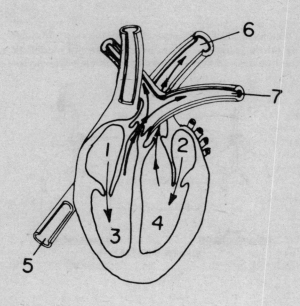

40. Which number indicates the structure which conducts blood to the lungs?
    - (1) 1
    - (2) 5
    - (3) 6
    - (4) 7

41. Which number indicates the receiving chamber for blood returning from a kidney?
    - (1) 1
    - (2) 2
    - (3) 3
    - (4) 4

42. The last heart chamber the blood leaves before moving toward capillaries in the arm is indicated by number
    - (1) 1
    - (2) 2
    - (3) 3
    - (4) 4

43. Blood entering vessel 5 would be expected to have more
    - (1) O_2 than in chamber 1
    - (2) O_2 than in vessel 6
    - (3) CO_2 than in chamber 3
    - (4) CO_2 than in vessel 6

44. Examination of the walls of vessels 5 and 6 would reveal
    - (1) more cardiac muscle in vessel 5 than in vessel 6
    - (2) less smooth muscle in vessel 5 than in vessel 6
    - (3) equal amounts of smooth muscle in both vessels
    - (4) less voluntary muscle in vessel 6 than in vessel 5

Directions: Base your answers to questions 45 through 47 on the diagrams below, which represent models of the human respiratory system, and on your knowledge of biology.

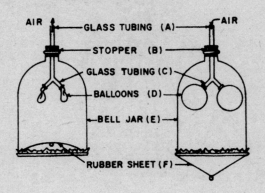

45. The rubber sheet in this model most likely represents the human
    (1) ribs
    (2) chest
    (3) lungs
    (4) diaphragm

46. Bronchitis, an inflammation or irritation of the bronchi in humans, may be expected to occur within a human structure represented in the diagram by
    (1) F
    (2) B
    (3) C
    (4) E

47. The rib cage is represented by structure
    (1) E
    (2) F
    (3) C
    (4) D

48. In humans, which substance is directly responsible for controlling the calcium levels of the blood?
    (1) adrenalin
    (2) insulin
    (3) parathormone
    (4) thyroxin

49. The bones of the lower arm are connected to the muscles of the upper arm by
    (1) ligaments
    (2) tendons
    (3) cartilage
    (4) skin

50. Glands located within the digestive tube include
    (1) gastric glands and thyroid glands
    (2) gastric glands and intestinal glands
    (3) thyroid glands and intestinal glands
    (4) adrenal glands and intestinal glands

Directions (51–53): For each statement in questions 51 through 53, select the structure, chosen from the list below, which is best described by that statement.

Structures

(1) Cardiac muscle
(2) Smooth muscle
(3) Voluntary muscle
(4) Ligament
(5) Tendon

51. This tissue is directly involved in the peristaltic activity of the alimentary canal.

52. This structure binds the ends of bones together.

53. This tissue could be classified as either an extensor or a flexor.

54. Compared to an endocrine response, a nerve response differs in that the nerve response generally
    (1) aids in homeostasis
    (2) requires a chemical secretion
    (3) is longer in duration
    (4) is more rapid

55. Anaerobic respiration is considered to be less efficient than aerobic respiration because
    (1) less lactic acid is formed during anaerobic respiration than aerobic respiration
    (2) anaerobic respiration requires more oxygen than aerobic respiration
    (3) the net gain of ATP molecules is less in anaerobic respiration than in aerobic respiration
    (4) less energy is required during anaerobic respiration than aerobic respiration

56. Which organism ingests food by engulfing it with pseudopods?
    (1) grasshopper   (3) earthworm
    (2) Paramecium   (4) Ameba

Directions (57–61): For each phrase in questions 57 through 61, select the number of the structure of the human central nervous system, chosen from the list below, that is best described in the phrase.

Structures

(1) Cerebrum
(2) Cerebellum
(3) Medulla
(4) Spinal cord

57. Interprets sensory impulses

58. Regulates peristalsis

59. Functions as a memory center

60. Conducts impulses between the brain and other body structures

61. Coordinates muscles and bones used in walking and running

62. Which type of vessel normally contains valves that prevent the backward flow of materials?
    (1) artery
    (2) arteriole
    (3) capillary
    (4) vein

63. Homeostatic regulation of the body is made possible through coordination of all body systems. This coordination is achieved mainly by
    (1) respiratory and reproductive systems
    (2) skeletal and excretory systems
    (3) nervous and endocrine systems
    (4) circulatory and digestive systems

64. In Paramecia, nitrogenous wastes are removed from the cell through
    (1) Malpighian tubules
    (2) cell membranes
    (3) tracheal tubes
    (4) food vacuoles

65. Which organism has a respiratory system that contains specialized oxygen-conducting structures?
    (1) Ameba
    (2) Hydra
    (3) Paramecium
    (4) grasshopper

Directions: Base your answers to questions 66 through 68 on your knowledge of biology and the diagram below which shows a human heart with the blood flow indicated by arrows.

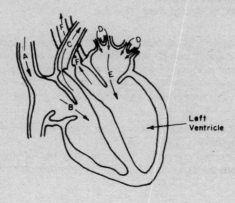

66. Oxygen-rich blood is present at locations
    (1) A and C
    (2) B amd E
    (3) C and F
    (4) D and E

67. The blood is flowing through veins at points
    (1) A and C
    (2) A and F
    (3) A and D
    (4) C and F

68. The blood flowing at point C is on its way to the
    (1) liver
    (2) head
    (3) lungs
    (4) kidneys

69. Which is a product formed during the anaerobic phase of cellular respiration?
    (1) pyruvic acid
    (2) water
    (3) oxygen
    (4) glucose

70. A synapse is described as the
    (1) electrochemical message of the nervous system
    (2) functional unit of the nervous system
    (3) change in the environment that brings about a response
    (4) space between adjacent neurons

71. The aerobic respiration of a molecule of glucose releases more energy than the anaerobic respiration of a molecule of glucose because, in aerobic respiration,
    (1) carbon dioxide is used
    (2) more chemical bonds are broken
    (3) oxygen is released
    (4) lactic acid is formed

72. Which activity is an example of the life process known as synthesis?
    (1) An organic compound is broken down and energy is released.
    (2) Starch is formed by the chemical bonding of glucose molecules.
    (3) A large molecule is broken down into smaller molecules.
    (4) Oxygen moves into a cell through the cell membrane.

73. In animals, the organelles in which aerobic cellular respiration occurs are known as
    (1) ribosomes
    (2) chloroplasts
    (3) nuclear membranes
    (4) mitochondria

74. The hydrolysis of maltose is catalyzed by
    (1) glucose
    (2) water
    (3) maltase
    (4) protease

75. An enzyme that acts as a catalyst in the digestion of certain carbohydrates is
    (1) maltase
    (2) lipase
    (3) protease
    (4) ATP-ase

76. An organism with a one-way digestive tube is the
    (1) Paramecium
    (2) earthworm
    (3) Ameba
    (4) Hydra

77. Exchange of oxygen and carbon dioxide between the external environment and the blood occurs in the
    (1) pharynx
    (2) alveoli
    (3) trachea
    (4) bronchi

78. Which organism moves by the interaction of muscles and chitinous appendages?
    (1) Hydra
    (2) Paramecium
    (3) grasshopper
    (4) human

79. Which is an activity controlled primarily by the automatic nervous system?
    (1) thinking during an exam
    (2) writing your name
    (3) regulating heartbeat
    (4) chewing food

80. Which organism is essentially a sessile animal?
    (1) Ameba
    (2) grasshopper
    (3) earthworm
    (4) Hydra

81. Which substance is secreted by a nerve cell but not by a muscle cell?
    (1) glycerol
    (2) neurohumor
    (3) lactic acid
    (4) uric acid

82. Animals commonly store energy in the form of
    (1) fat and glycogen
    (2) waxes and oils
    (3) minerals and urea
    (4) water and carbon dioxide

83. Which human excretory organ breaks down red blood cells and synthesizes urea?
    (1) lung
    (2) kidney
    (3) skin
    (4) liver

84. In humans, the center for regulating the amount of oxygen in the blood is situated in the
    (1) cerebrum
    (2) cerebellum
    (3) medulla
    (4) spinal cord

85. The synthesis of fibrin is controlled directly by
    (1) enzymes present in the blood
    (2) hormones secreted by the pituitary gland
    (3) neurons in the medulla of the brain
    (4) phagocytic cells within the intercellular fluid

86. What is the main function of the mucus secreted by cells of the external body wall of the earthworm?
    (1) It acts as a stimulus upon the paired nephridia.
    (2) It permits transpiration to be accomplished.
    (3) It is an excretory waste product.
    (4) It provides a moist surface for gas exchange.

87. Two end-products of aerobic respiration are
    (1) oxygen and alcohol
    (2) oxygen and water
    (3) carbon dioxide and water
    (4) carbon dioxide and oxygen

88. Which animal excretes wastes by the action of nephridia?
    (1) Paramecium
    (2) Hydra
    (3) earthworm
    (4) grasshopper

89. The circulatory system of the earthworm is most similar in structure and function to that of a
    (1) Hydra
    (2) protozoan
    (3) grasshopper
    (4) human

90. Freshwater unicellular organisms excrete their nitrogenous wastes mainly in the form of
    (1) ammonia
    (2) urine
    (3) urea
    (4) uric acid

91. Which organism possesses a single opening which functions as a mouth for the ingestion of food and also as an anus for the elimination of undigested materials?
    (1) human
    (2) grasshopper
    (3) Hydra
    (4) earthworm

92. Which substances, when found within the food vacuole of an Ameba, indicate that this organelle has a digestive function?
    (1) hydrolytic enzymes
    (2) oxygen molecules
    (3) inorganic salts
    (4) carbon dioxide molecules

93. A central nervous system is present in
    (1) Hydrae and humans
    (2) Paramecia and Hydrae
    (3) earthworms and grasshoppers
    (4) Amebae and humans

94. The body normally responds to low concentrations of sugar in the blood by secreting
    (1) glucagon
    (2) estrogen
    (3) insulin
    (4) testosterone

95. The enzyme salivary amylase will act on starch, but not on protein. This action illustrates that salivary amylase
    (1) contains starch
    (2) is not reusable
    (3) is chemically specific
    (4) lacks protein

96. In the grasshopper, the structures which are associated with the removal of gases from the organism are known as
    (1) Malpighian tubules
    (2) nephridia
    (3) lungs
    (4) tracheae

97. Which activity is an adaptation that enables an earthworm to live on land?
    (1) secretion of mucus which moistens the skin
    (2) production of nitrogenous waste products
    (3) oxidation of glucose with the aid of enzymes
    (4) digestion of food sequentially in a one-way tract

98. Organisms make energy readily available by transferring the chemical bond energy of organic molecules to
    (1) mineral salts
    (2) adenosine triphosphate
    (3) light energy
    (4) nitrogenous wastes

99. How are nutrients transported from the blood of an earthworm to the muscle cells of its body wall?
    (1) as a result of blood flowing directly into muscle cells
    (2) as a result of diffusion through capillary walls
    (3) through the pores at the ends of nephridia
    (4) through the skin from the outside environment

Directions: Base your answers to questions 100 through 104 on your knowledge of biology and on the graph below which shows the extent to which carbohydrates, proteins, and fats are chemically digested as food passes through the human digestive tract. The letters represent sequential structures that make up the digestive tract.

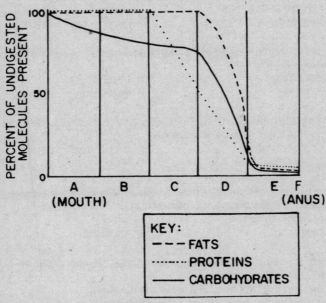

100. Proteins are digested in both
    (1) *A* and *B*
    (2) *B* and *C*
    (3) *C* and *D*
    (4) *A* and *C*

101. The organ represented by letter *C* is most probably the
    (1) esophagus
    (2) stomach
    (3) small intestine
    (4) large intestine

102. Enzymes secreted by the pancreas enter the system at
    (1) *E*
    (2) *B*
    (3) *C*
    (4) *D*

103. The final products of digestion are absorbed almost entirely in
    (1) *F*
    (2) *B*
    (3) *C*
    (4) *D*

104. Water is removed from the undigested material in
    (1) A   (3) E
    (2) B   (4) D

105. In an Ameba, the structures formed as a result of the ingestion of yeast cells are known as
    (1) food vacuoles   (3) cell walls
    (2) contractile vacuoles   (4) buds

106. An increase in the concentration of ATP in a muscle cell is a direct result of which life function?
    (1) respiration   (3) digestion
    (2) reproduction   (4) excretion

107. The functional unit of the human kidney is known as a
    (1) nephridium   (3) nephron
    (2) Malpighian tubule   (4) urinary bladder

108. In the human body, the blood with the greatest concentration of oxygen is found in the
    (1) left atrium of the heart   (3) nephrons of the kidney
    (2) cerebrum of the brain   (4) lining of the intestine

Directions (109–113): For each phrase in questions 109 through 113, select the human structure, *chosen from the list below*, that is best described by that phrase.

Human Structures

(1) Bones
(2) Cartilage tissues
(3) Ligaments
(4) Smooth muscles
(5) Tendons
(6) Voluntary muscles

109. Cause peristalsis in the alimentary canal

110. Serve as extensors and flexors

111. Serve as levers for body movement

112. Bind the ends of bones together

113. Attach the muscles to bones

114. Which correctly matches an organism with the structures it uses for locomotion?
    (1) grasshopper—chitinous appendages and setae
    (2) human—muscles and bone
    (3) Hydra—flagella and pseudopodia
    (4) Paramecium—pseudopodia and muscles

115. During aerobic respiration, which atoms are removed from glucose molecules by the action of enzymes?
    (1) nitrogen   (3) hydrogen
    (2) iron   (4) calcium

116. The diagram below represents a portion of onion epidermal tissue as seen through a compound microscope. Which observation would probably be made if the water surrounding the cells were replaced by a concentrated salt solution?

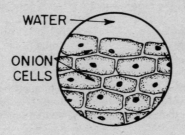

WATER
ONION CELLS

    (1) The walls surrounding each of the onion cells would turn blue-black.
    (2) Large spaces would appear between the cell membranes and cell walls.
    (3) The tissue would break down into individual cells.
    (4) The nucleoli inside the nuclei would become larger.

117. In freshwater protozoa, most excess water is eliminated by the process of
    (1) cyclosis   (3) active transport
    (2) hydrolysis   (4) pinocytosis

118. The process by which unicellular organisms ingest large organic molecules without previously digesting them is known as
    (1) diffusion   (3) osmosis
    (2) passive transport   (4) pinocytosis

Answers

1. 2	31. 2	61. 2	91. 3
2. 3	32. 1	62. 4	92. 1
3. 3	33. 4	63. 3	93. 3
4. 3	34. 1	64. 2	94. 1
5. 3	35. 4	65. 4	95. 3
6. 1	36. 2	66. 4	96. 4
7. 2	37. 2	67. 3	97. 1
8. 4	38. 1	68. 3	98. 2
9. 4	39. 4	69. 1	99. 2
10. 3	40. 4	70. 4	100. 3
11. 3	41. 1	71. 2	101. 2
12. 1	42. 4	72. 2	102. 4
13. 4	43. 4	73. 4	103. 4
14. 2	44. 2	74. 3	104. 3
15. 3	45. 4	75. 1	105. 1
16. 4	46. 3	76. 2	106. 1
17. 3	47. 1	77. 2	107. 3
18. 2	48. 3	78. 3	108. 1
19. 2	49. 2	79. 3	109. 4
20. 2	50. 2	80. 4	110. 6
21. 3	51. 2	81. 2	111. 1
22. 1	52. 4	82. 1	112. 3
23. 4	53. 3	83. 4	113. 5
24. 4	54. 4	84. 3	114. 2
25. 2	55. 3	85. 1	115. 3
26. 4	56. 4	86. 4	116. 2
27. 1	57. 1	87. 3	117. 3
28. 3	58. 3	88. 3	118. 4
29. 4	59. 1	89. 4	
30. 4	60. 4	90. 1	

Explanatory Answers

1. **(2)** In the process of synthesis, humans build up their own specific proteins from amino acids ingested. INCORRECT CHOICES: (1) Regulation is the coordination exercised by an organism to maintain stability in a

changing environment. (3) Respiration is the release of energy from food. (4) Excretion is the process of getting rid of wastes.

2. **(3)** The white blood cells' function is to protect the body by the engulfment (phagocytosis) of foreign antigens. INCORRECT CHOICES: (1) The red blood cells' function is to transport oxygen. (2) The platelets' function is blood clotting. (4) Pathogens are disease-causing organisms.

3. **(3)** Oxygen, through diffusion, enters body cells due to passive transport. INCORRECT CHOICES: (1) The higher concentration of oxygen is in the capillaries, not the cell. (2) Osmosis is the diffusion of water. (4) Pinocytosis is the engulfment by pinocytic vesicles of food too large to pass through the cell membrance.

4. **(3)** When oxygen is deficient or absent some cells respire anaerobically producing lactic acid and a relatively small amount of ATP. INCORRECT CHOICES: (1) Hydrolysis involves the breakdown of large organic molecules to their building blocks. (2) Dehydration synthesis involves the building up of large organic molecules from their building blocks. (4) Aerobic respiration of glucose leads to the production of carbon dioxide, water, and a relatively large amount of ATP.

5. **(3)** Amino acids are the end products of protein digestion and simple sugars are the end products of carbohydrate digestion. INCORRECT CHOICES: (1) Bile is the digestive juice secreted by the liver; enzymes are found in the other digestive juices. (2) Both proteins and lipids consist of large molecules which must be digested to small, soluble molecules so they may be distributed throughout the body. (4) Both proteins and carbohydrates consist of large molecules which must be digested.

6. **(1)** The cell membrane is a complex, porous structure composed mainly of protein and lipid materials. INCORRECT CHOICES: (2) Sugars and starches are not found in large quantities in the cell membrane. (3) Cellulose makes up the cell walls of plant cells. (4) Polysaccharides such as starches and cellulose are not found in cell membranes.

7. **(2)** The lungs excrete carbon dioxide and water vapor while the kidneys excrete water, salts, and nitrogenous wastes. INCORRECT CHOICES: (1) The heart pumps blood throughout the body including the skin which excretes water and salts as sweat. (3) The liver discharges bile salts into the small intestine while the testes are the male reproductive organs forming sperm and secreting male sex hormones. (4) The kidneys are excretory organs, but the pancreas is not; the pancreas secretes digestive juices into the small intestine and hormones into the blood.

8. **(4)** Endocrine glands secrete hormones, which control the activities of other cells, into the blood. INCORRECT CHOICES: (1) Enzymes are organic catalysts which control the rates of chemical reactions in cells. (2) Neurohumors are secreted by the ends of neurons and permit impulses to cross synapses. (3) Auxins are plant growth hormones.

9. **(4)** Pinocytosis is a process in which large molecules that are too large to pass through the cell membrane may be engulfed and brought within the cell. INCORRECT CHOICES: (1) Passive transport is any form of transport in which cells do not contribute energy for the movement of materials. (2) Osmosis, a form of passive transport, is the diffusion of water into and out of cells. (3) Cytoplasmic streaming is a means of circulation of materials within cells.

10. **(3)** In aerobic respiration, oxygen serves as the final acceptor for hydrogen released from organic molecules, such as glucose. INCORRECT CHOICES: (1) Pyruvic acid is formed in the anaerobic phase of respiration. (2) Glucose is broken down in aerobic respiration to carbon dioxide and water. (4) Hydrolysis changes proteins to amino acids.

11. **(3)** The diagram is of the nephron of the kidney. Filtering of the blood begins in the glomerulus and passes into the Bowman's capsule pictured in number 3. INCORRECT CHOICES: (1) Number 5 is pointing to the convoluted tubule which reabsorbes much of the filtrate. (2) Number 6 is pointing to the renal vein which transports the filtered blood back toward the heart. (4) This number 4 is the collecting tubule which holds the waste products and leads to the pelvis of the kidney for excretion.

12. **(1)** Excess water which was filtered out of the blood in the Bowman's capsule is now reabsorbed here at the tubule. INCORRECT CHOICES: Refer to number 11 above.

13. **(4)** The urine collects after reabsorption, in the collecting tubule, pictured in number 4. INCORRECT CHOICES: Refer to number 11 above.

14. **(2)** The digestive tube is a self-contained tube beginning with the mouth and ending with the anus. Accessory glands which aid in digestion are the liver and pancreas. INCORRECT CHOICES: (1) The liver is outside the tube and the stomach is part of the tube. (3) The pancreas is an accessory gland and the small intestine is a large part of the tube. (4) The pancreas again is an accessory gland while the stomach is the storage aspect of the tube.

15. **(3)** The grasshopper has an internal, open circulatory system which

transports materials. INCORRECT CHOICES: (1) The Hydra does not have a circulatory system. (2) and (4) The earthworm and human have closed circulatory systems.

16. **(4)** The term transport simply means to obtain and distribute materials. INCORRECT CHOICES: (1) Excretion is the life process aimed at the removal of metabolic wastes. (2) Locomotion is the means by which an organism obtains mobility. (3) Synthesis is the manufacture of complex molecules from simple compounds and/or elements.

17. **(3)** The legs contain voluntary muscle tissue used in walking and running. INCORRECT CHOICES: (1) The brain consists mostly of nerve tissue; a small amount of involuntary muscle is found in the walls of arteries in the brain. (2) The heart consists of involuntary cardiac muscle tissue. (3) The small intestine contains involuntary muscle in its walls.

18. **(2)** The earthworm is a heterotroph because it lacks the ability to synthesize nutrients from inorganic raw materials and must consume preformed organic compounds. INCORRECT CHOICES: (1) The earthworm lacks the ability to synthesize nutrients from inorganic raw materials. (3) All living things synthesize proteins. (4) The earthworm cannot fix nitrogen from the soil.

19. **(2)** Locomotion provides for greater opportunity to seek out and ingest a variety of nutrients, while sessile organisms must rely on a variety of means such as photosynthesis, or water currents. INCORRECT CHOICES: (1) Circulation is an internal process of transport of ingest particles. (3) Synthesis is the manufacture of complex substances. (4) Excretion again is also an internal process which is concerned with the removal of metabolic wastes.

20. **(2)** Neurons are specialized nerve tissue which compose clusters called ganglia. INCORRECT CHOICES: (1) Smooth muscle cells compose the lining of the digestive system. (3) Phagocytes are white blood cells which act in the body's defense. (4) Striated muscle cells compose the muscle tissue of large muscles of movement.

21. **(3)** Neurohumors can be compared to hormones in their activities which amount to the stimulation of a neuron to allow or retard the passage of an impulse. INCORRECT CHOICES: (1) Hormones are secreted by glands not by nerve cells. (2) Auxins are growth hormones in plants. (4) Toxins are poisons that may affect either plants or animals.

22. **(1)** Since the Hydra lives in a watery environment, it needs no specialized transport system and relies on the simple processes listed. INCORRECT

CHOICES: (2)Transpiration is a water removal process in plants, cyclosis is the streaming motion of the cytoplasm, and pinocytosis is the engulfment of particles too large to pass through the cell membrane. (3) and (4) This system is indicative of a land dweller and the Hydra lives in the water.

23. **(4)** The esophagus is composed of longitudinal and circular muscles which alternately contract and expand giving the peristaltic effect. INCORRECT CHOICES: (1) The term peristaltic motion refers to the movement of food through the digestive cavity. The liver is concerned with the production of bile. (2) The pancreas is concerned with the production of pancreatin enzymes. (3) The oral cavity serves for mechanical and chemical digestion of food as well as for ingestion.

24. **(4)** A motor neuron delivers neural impulses to muscles or glands to elicit a response to the original stimuli. INCORRECT CHOICES: (1) The associative neuron transmits the impulses from the spinal cord to the brain. (2) The receptors are the sense organs. (3) The sensory neuron transmits impulses from the sense organ to the central nervous system.

25. **(2)** The term absorption means the diffusion of water and dissolved materials into cells. Therefore the movement of end products diffusing through membranes and eventually into body fluids constitutes absorption. INCORRECT CHOICES: (1) Osmosis applies only to the movement of water. (3) Synthesis means to manufacture complete substances from simple ones. (4) Photosynthesis is the food-making process of autotrophs.

26. **(4)** Both longitudinal and circular muscles contract and expand, inching the worm along the earth's surface. The setae act as anchors. INCORRECT CHOICES: (1) The flagella is an organelle of locomotion in the Euglena. (2) Cilia are organelles of locomotion in Paramecium. (3) Since the earthworm does not have a skeletal structure, bones could not be an answer.

27. **(1)** As an invertebrate example, both the grasshopper and earthworm have the nervous systems listed. However, the earthworm is the only one that appears as an answer. INCORRECT CHOICES: (2) The Hydra possesses a nerve net; brain and nerve cord are absent. (3) The Ameba's regulation is by chemical means only. (4) The human has an anterior brain and dorsal nerve cord.

28. **(3)** Glandular secretions (specifically endocrine) regulate metabolism of the organism. These secretions are called hormones. INCORRECT CHOICES: (1) A nucleic acid controls the transmission and maintenance of heredity. (2) A nitrogenous base is a component for a nucleic acid. (4) A neu-

rohumor is secreted at the end of nerve fibers at the synapse to accelerate or retard the impulse.

29. **(4)** Uric acid is produced by the grasshopper in a water conservation move. This is necessary in that this organism lives entirely on land. INCORRECT CHOICES: (1) The Paramecium, a water-dwelling organism, excretes ammonia directly into the water which dilutes it. (2) and (3) The Hydra and the Ameba fall into the same category as the Paramecium.

30. **(4)** Hydrolytic enzymes function in digestion. Both autotrophs and heterotrophs carry on digestion. INCORRECT CHOICES: (1), (2), and (3) As stated above all organisms carry on digestion and need hydrolytic enzymes to carry out the process.

31. **(2)** The ductless glands are the endocrines and they secrete hormones. INCORRECT CHOICES: (1) Enzymes are protein in nature and therefore produce ribosomes. (3) Lachrymal fluids are produced by the eyes and are better known as tears. (4) Excretory fluids are produced in various excretory glands in the different organisms.

32. **(1)** The Hydra ingests and egests material through its only pore, the mouth. Therefore it is said that the hydra has a "two-way digestive system." INCORRECT CHOICES: (2) The earthworm, having a mouth and anus represents a "one-way" digestive system. (3) The grasshopper also has a mouth and anus and represents a one-way system. (4) The human represents the most complex one-way digestive system.

33. **(4)** Stimuli are changes in an organism's surroundings that cause it to respond. INCORRECT CHOICES: (1) Effectors, muscles or glands, produce the response. (2) Receptors, or sense organs, receive stimuli. (3) An impulse is an electrochemical message which travels down a neuron.

34. **(1)** The human nervous system secretes neurohumors at the end of the synapses and the endocrine glands secrete hormones directly into the bloodstream which are then transported to target organs. INCORRECT CHOICES: (2) The nervous system and endocrine system rates of response are entirely different. (3)Neural responses may last seconds or parts of seconds while an endocrine response may last indefinitely. (4) Hormones travel only through the bloodstream.

35. **(4)** An external respiratory system is characteristic of organisms such as the Paramecium and Hydra that live in the water, and of organisms such as the earthworm that live in moist environments. INCORRECT CHOICES: (1) Only the Hydra is a sessile organism. (2) None of the choices are

photosynthetic. (3) Only the earthworm has a closed circulatory system. The other two have no specialized system for circulation.

36. **(2)** Animals do not make food, a form of preformed organic compounds; therefore, they must take in organic materials which are already formed. INCORRECT CHOICES: (1) Carbon dioxide and water are both inorganic. (3) Hydrolysis is a breaking down, not a synthesis of substances. (4) Oxidation results in a release of energy.

37. **(2)** The nephrons are tiny microscopic tubules within the kidney that filter the blood and reabsorb valuable water and minerals. INCORRECT CHOICES: (1) The alveoli are the tiny sacs within the lungs where gas exchange occurs. (3) A sweat gland is located in the epidermal layer as a pore and the dermal layer of skin as a gland. (4) The liver, located below the diaphragm, filters the blood and stores glycogen.

38. **(1)** The alveolus is the site for gas exchange. Carbon dioxide and water vapor exit the bloodstream and oxygen enters it. INCORRECT CHOICES: Refer to number 37 above.

39. **(4)** Old red blood cells are broken down into waste products and those wastes are excreted into the bile. This occurs in the liver. INCORRECT CHOICES: Refer to number 37 above.

40. **(4)** Number 7 is the pulmonary artery which carries blood from the heart to the lungs. INCORRECT CHOICES: (1) Number 1 represents the right atrium. (2) Number 5 depicts the vena cava. (3) Number 6 indicates the aorta which carries blood from the heart to all parts of the body.

41. **(1)** Blood returning from all parts of the body enters the right atrium. INCORRECT CHOICES: (2) Number 2 represents the left atrium which receives blood returning from the lungs. (3) Number 3 is the right ventricle which pumps blood to the lungs. (4) Number 4 depicts the left ventricle which pumps blood to the body.

42. **(4)** The left ventricle pumps blood from the heart to all parts of the body. INCORRECT CHOICES: (1) The right atrium receives blood from the body. (2) The left atrium receives blood from the lungs. (3) The right ventricle pumps blood to the lungs.

43. **(4)** Blood leaving the heart through the aorta (6) has less CO_2 than blood entering through the vena cava (5); CO_2 is lost as blood passes through the lungs. INCORRECT CHOICES: (1) Blood in the right atrium (1) and vena cava (5) are both poor in oxygen. (2) Blood in the vena cava has *less*

oxygen than blood in the aorta because oxygen was picked up in the lungs. (3) Blood in the vena cava has the same amount of CO_2 as blood in the right ventricle.

44. **(2)** A vein such as the vena cava has less muscular walls than an artery such as the aorta. INCORRECT CHOICES: (1) Cardiac muscle is found in the heart and not in blood vessels. (3) Arteries have more smooth muscle than veins do. (4) Voluntary muscle is found in the skeletal muscles not in blood vessels.

45. **(4)** The rubber sheet in the diagram is shown relaxing, thereby expelling air from the jar, and contracting, thereby drawing air into the jar. The human diaphragm acts in the same way. INCORRECT CHOICES: (1) The ribs protect the heart and lung tissues. (2) The chest, as a broad term, includes the ribs, heart, and lungs. (3) The lungs are epithelial tissue which accept air drawn into them by the diaphragm.

46. **(3)** Letter C represents the bronchi or bronchial tubes where bronchitis would occur. INCORRECT CHOICES: (1) Letter F represents the diaphragm. (2) Letter B may represent the larnyx. (4) Letter E represents the chest cavity wall.

47. **(1)** The rib cage or outer perimeter of the lung is Letter E. INCORRECT CHOICES: Refer to number 46 above.

48. **(3)** Parathormone which is secreted by the parathyroids located in the proximity of the thyroid gland, controls calcium levels in the body. Improper functioning of the parathyroids results in tetany. INCORRECT CHOICES: (1) Adrenalin, the "fight or flight" hormone is secreted by the adrenal glands and controls the oxidation of glucose. (2) Insulin maintains the proper glucose levels in the blood. (4) Thyroxin, which contains large amounts of iodine, controls the general metabolic rate in the body.

49. **(2)** Tendons are tough connective tissues that do attach muscles to bones. INCORRECT CHOICES: (1) Ligaments attach bones to bones. (3) Cartilage is a strong but pliable support tissue. (4) The skin, epidermal tissue, covers the exterior of the body and functions to maintain body temperature and to prevent infectious organisms from entering the underlying tissue.

50. **(2)** The term gastric means within the stomach, and intestinal, of course, deals with the intestinal tract. INCORRECT CHOICES: (1) Gastric is correct but thyroid glands are located in the neck near the larynx. (3) Again the term that is incorrect for the above reason is thyroid. (4) The adrenal glands sit atop the kidneys.

51. **(2)** Smooth muscle is involuntary muscle which controls activities such as the movement of food through the alimentary canal (peristalsis). INCORRECT CHOICES: (1) Cardiac muscle is the type of muscle found in the heart. (5) Tendons are connective tissues which attach muscles to bones.

52. **(4)** Ligaments are tough tissues which connect bones to bones at movable joints. INCORRECT CHOICES: See numbers 51 and 53.

53. **(3)** A flexor is a voluntary, skeletal muscle which bends a joint; an extensor straightens a joint. INCORRECT CHOICES: See numbers 51 and 52.

54. **(4)** Nerve impulses are electrochemical in nature and travel at great speeds. INCORRECT CHOICES: (1) Both nervous and endocrine controls contribute to a constant internal environment. (2) Neurons secrete neurohumors; endocrine glands secrete hormones. (3) Hormones, secreted by endocrine glands, travel through the blood and have a lasting effect.

55. **(3)** Quantitatively, only 2 molecules of ATP are released in anaerobic respiration whereas 38 molecules of ATP are released in aerobic respiration. INCORRECT CHOICES: (1) Lactic acid is not produced in aerobic respiration. (2) Anaerobic means in the absence of oxygen. (4) Activation energy of 2 ATP molecules is necessary for both reactions to begin. Less energy is released, not required, during anaerobic respiration.

56. **(4)** The pseudopods are organelles of ingestion and locomotion in the Ameba. INCORRECT CHOICES: (1) The grasshopper has specialized mouthparts and legs for the ingestion of food. (2) The Paramecium has a specialized oral groove and cilia for ingestion. (3) The earthworm has a fleshy proboscis and a muscular pharynx for ingestion.

57. **(1)** Sensory impulses become sensations in the cerebrum. INCORRECT CHOICES: (2) The cerebellum coordinates motor activity. (3) The medulla controls involuntary body activities. (4) The spinal cord is a center for many reflex actions and conducts impulses between the brain and other parts of the body.

58. **(3)** The medulla controls involuntary body processes such as peristalsis. INCORRECT CHOICES: (1) The cerebrum interprets sensory impulses, initiates motor activity, and is the seat of consciousness and memory. (3) and (4) See number 57 above.

59. **(1)** The cerebrum functions as a memory center. INCORRECT CHOICES: (2), (3), and (4) See number 57 above.

60. **(4)** The spinal cord conducts impulses between the brain and other body structures. INCORRECT CHOICES: (1), (2), and (3) The cerebrum, cerebellum, and medulla are parts of the brain.

61. **(2)** The cerebellum coordinates muscles and bones used in walking and running. INCORRECT CHOICES: (1) The cerebrum initiates activities such as walking and running. (3) The medulla controls involuntary body processes. (4) Impulses to the muscles used in walking and running are conducted to them from the brain through the spinal cord.

62. **(4)** Veins have no muscular tissue for the pumping of blood toward the heart. They in turn rely on the constrictions of muscles and valves to push the blood to the heart. INCORRECT CHOICES: (1) An artery has muscles in its walls to aid the flow of blood. (2) An arteriole, or small artery, like the artery has muscles. (3) A capillary, the smallest of the blood tubes, delivers blood cells to body cells.

63. **(3)** The nervous system coordinates stimuli and responses through a network of nerve fibers or neurons. The endocrine system in response to the stimuli delivered by the neurons secretes hormones which act in countless ways to balance the body's systems. INCORRECT CHOICES: (1) Osmosis, diffusion, and active transport only regulate the homeostatic balance of water and dissolved minerals. (2) The skeletal system coordinates movement and provides support, and the excretory system provides for the removal of metabolic wastes. (4) The circulatory system provides for the transport of oxygen and carbon dioxide plus the products of digestion provided by the digestive system.

64. **(2)** In Paramecia, nitrogenous wastes diffuse out of the cell through the cell membrane. INCORRECT CHOICES: (1) Malpighian tubules are excretory tubules which remove nitrogenous wastes from the blood of the grasshopper. (3) Insects such as the grasshopper have a system of tracheal tubules for the intake, distribution, and removal of respiratory gases. (4) In Paramecia, ingested food materials accumulate in food vacuoles where digestion occurs.

65. **(4)** The tracheal tubes in the grasshopper carry oxygen-rich air to all tissues. INCORRECT CHOICES: (1) Oxygen dissolved in the water surrounding the Ameba diffuses through the cell membrane. (2) Oxygen diffuses into the Hydra from the water surrounding the gastrovascular cavity. (3) Oxygen passes by osmosis through the cell membrane of the Paramecium.

66. **(4)** Oxygen-rich blood enters the heart from the lungs through the pulmonary veins (*D*) into the left auricle (*E*). INCORRECT CHOICES: (1) Oxygen-poor blood enters the heart from the body through the vena cava (*A*)

and is pumped to the lungs through the pulmonary artery (C). (2) The blood in the right auricle (B) is oxygen-poor while the blood in the left auricle (E) is oxygen-rich. (3) Oxygen-poor blood is present in the pulmonary artery (C) to the lungs while the blood in the aorta (F) is oxygen-rich.

67. **(3)** Blood is flowing toward the heart in the vena cava (A) and the pulmonary veins (D). INCORRECT CHOICES: (1) Blood flows away from the heart in the pulmonary artery (C). (2) Blood flows away from the heart in the aorta (F). (4) The pulmonary artery (C) and the aorta (F) are both arteries.

68. **(3)** The pulmonary artery (C) carries blood to the lungs. INCORRECT CHOICES: (1), (2), and (4) The aorta (F) carries blood to the liver, the head, and the kidneys.

69. **(1)** Glucose is broken down into two molecules of pyruvic acid during the anaerobic phase of cellular respiration. INCORRECT CHOICES: (2) Water is formed during the aerobic phase of cellular respiration. (3) Oxygen is used as a hydrogen acceptor during the aerobic phase of cellular respiration. (4) Glucose is used during the anaerobic phase of cellular respiration.

70. **(4)** Synapses are junction spaces between adjacent neurons. INCORRECT CHOICES: (1) The electrochemical message of the nervous system is known as the nerve impulse. (2) The functional unit of the nervous system is the neuron. (3) A change in the environment that brings about a response is known as a stimulus.

71. **(2)** Compared to anaerobic respiration, more energy storing bonds are broken down in aerobic respiration. INCORRECT CHOICES: (1) Carbon dioxide is one product of aerobic respiration. (3) Oxygen is used, not released, in aerobic respiration. (4) Lactic acid is formed during anaerobic respiration.

72. **(2)** The term synthesis refers to the manufacture of more complex substances through dehydration. This choice is the only one which meets these criteria. INCORRECT CHOICES: (1) Since synthesis is the building process and this choice uses the term "broken down" we would have to eliminate it. (3) Again the term "broken down" is used and this of course is not synthesis. (4) This choice refers to diffusion which is transport.

73. **(4)** The mitochondria are the energy producing organelles of the cell. Aerobic respiration releases 38 molecules of ATP and therefore it would

stand to reason that an energy releasing process would occur in the energy producing organelle. INCORRECT CHOICES: (1) Ribosomes manufacture proteins. (2) Chloroplasts, found only in plant cells, are concerned with food production. (3) The nuclear membrane forms an interface between the nucleoplasm and the cytoplasm.

74. (3) Maltase is an enzyme which acts on maltose; -ase endings are characteristic of enzymes and the stem indicates the substrate. INCORRECT CHOICES: (1) Glucose is a simple sugar. (2) Water is added in hydrolysis reactions. (3) Proteases are enzymes which digest proteins.

75. (1) Maltase acts on the double sugar, maltose, breaking it down into two molecules of glucose. INCORRECT CHOICES: (2) Lipase acts in the digestion of lipids. (3) Protease is an enzyme which digests proteins. (4) The enzyme, ATP-ase, acts on ATP.

76. (2) In the earthworm's digestive tract, food passes in one direction from mouth to anus. INCORRECT CHOICES: (1) and (3) The Paramecium and the Ameba are one celled; they digest food in a food vacuole. (4) The gastrovascular cavity of the Hydra has one opening which serves as a mouth and anus.

77. (2) Alveoli are tiny air sacs at the ends of the air passageways in the lungs; capillaries which surround the alveoli permit gas exchange by diffusion. INCORRECT CHOICES: (1) Air passes from the nasal passages through the pharynx into the larynx. (3) Air passes from the larynx into the trachea. (4) Air passes from the bronchi through the bronchial tubes into the alveoli.

78. (3) The grasshopper moves by the interaction of muscles with chitinous appendages (legs and wings). INCORRECT CHOICES: (1) The Hydra is essentially sessile but does have contractile fibers which permit some motion. (2) The Paramecium moves by means of cilia. (4) Locomotion in humans is accomplished by the interaction of the muscles and the bones of the skeleton.

79. (3) The autonomic nervous system controls involuntary responses of many internal organs, including the heart. INCORRECT CHOICES: (1) Thinking during an exam is a function of the cerebrum. (2) Writing your name is voluntary behavior controlled by the cerebrum. (4) Chewing food is voluntary behavior controlled by the cerebrum.

80. (4) The Hydra is essentially sessile because it tends to remain in a fixed position much of the time. INCORRECT CHOICES: (1) The Ameba moves by means of pseudopodia. (2) Locomotion in the grasshopper is made pos-

sible by the interaction of muscles with its legs and wings. (3) The earthworm moves as a result of the action of its muscles and bristles (setae).

81. **(2)** Neurohumors are secreted by the ends of neurons and stimulate impulse production at the beginning of adjacent neurons. INCORRECT CHOICES: (1) Glycerol is a 3-carbon alcohol used in the synthesis of many lipids. (3) Lactic acid is a compound which is produced as a result of anaerobic respiration in muscle cells. (4) Uric acid is a nitrogenous waste produced by some animals (grasshoppers) as a water-conserving mechanism.

82. **(1)** Fat and glycogen are the most common storage products in animals. INCORRECT CHOICES: (2) Waxes and oils are among the specialized secretions produced by animals. (3) Minerals have many specific metabolic functions in animals but are usually harmful in high concentrations; urea is a nitrogenous waste product. (4) Water and carbon dioxide are waste products of respiration in which potential energy is released from organic compounds.

83. **(4)** The liver breaks down red blood cells and synthesizes urea from the waste products of protein metabolism. INCORRECT CHOICES: (1) The lungs are the principal means of carbon dioxide excretion. (2) Urea synthesized in the liver is absorbed into the blood and excreted by the kidneys along with water and salts. (3) Sweat glands in the skin excrete water and salts.

84. **(3)** The medulla controls many involuntary body activities, such as breathing, heartbeat, and blood pressure, which have an effect on the amount of oxygen in the blood. INCORRECT CHOICES: (1) The cerebrum is involved with the interpretation of the senses, the initiation of voluntary motor activity, consciousness, and memory. (2) The cerebellum coordinates motor activity. (4) The spinal cord is a center for many reflex actions, and conducts impulses between the brain and other body structures.

85. **(1)** The synthesis of fibrin results from enzyme-regulated reactions among disintegrated blood platelets, blood proteins, and calcium ions in the plasma. INCORRECT CHOICES: (2) Hormones secreted by the pituitary gland influence the activity of other endocrine glands. (3) Neurons in the medulla of the brain are involved in the control of a variety of involuntary body activities but not blood clotting. (4) Phagocytic cells within the intercellular fluid engulf bacteria and other pathogens.

86. **(4)** A moist surface is needed for exchange of gases through a membrane. INCORRECT CHOICES: (1) The paired nephridia are used for excretion. (2)

Transpiration is the loss of excess water from leaves. (3) An excretory waste product is a substance which is usually harmful to the organism.

87. **(3)** Aerobic respiration occurs when oxygen combines with a substance and then the wastes, carbon dioxide and water are formed. INCORRECT CHOICES: (1) Oxygen is needed for aerobic respiration, and alcohol is a product of fermentation. (2) Oxygen combines with materials and is not a product in aerobic respiration. (4) Oxygen is not an end product of aerobic respiration.

88. **(3)** Each segment in the earthworm contains a pair of excretory tubules called nephridia. INCORRECT CHOICES: (1) The Paramecium excretes wastes through the cell membrane and contractile vacuoles. (2) The Hydra excretes wastes through the cell membrane. (4) The Malpighian tubules in the grasshopper excrete wastes.

89. **(4)** The earthworm and human have closed tubular circulatory systems through which a liquid is pumped. INCORRECT CHOICES: (1) The Hydra has neither vessels nor a pump. Its gastrovascular cavity is an open tube into which water flows in and out. (2) Protozoans do not have a vascular circulatory system. (3) The grasshopper has an open circulatory system where the liquids flow freely within its body cavity.

90. **(1)** Ammonia is the water soluble nitrogenous waste secreted by organisms living in fresh water. INCORRECT CHOICES: (2) Urine is a liquid waste composed mainly of urea, salts, and excess water. (3) Urea is a waste product produced from the breakdown of proteins. (4) Uric acid is a solid nitrogen waste produced by some animals to conserve water.

91. **(3)** The Hydra has a mouth into which the water in which it lives passes in and out of its gastrovascular cavity. INCORRECT CHOICES: (1) The human has a mouth for ingestion and an anus for elimination. (2) The grasshopper has an opening at its head for ingestion and eliminates undigested materials through its anus. (4) The earthworm has a stoma or opening at its anterior end for ingestion and an anus at its posterior end for elimination.

92. **(1)** Hydrolytic enzymes are chemicals which permit water to combine with food in digestion. INCORRECT CHOICES: (2) Oxygen molecules combine with organic molecules to release energy. (3) Inorganic salts are not food and would not be digested. (4) Carbon dioxide is a product of respiration, not digestion.

93. **(3)** Both earthworms and grasshoppers have central nerve cords and a

ganglion or brain which serves as a central nervous system. INCORRECT CHOICES: (1) The Hydra has a nerve net, not a central nervous system. (2) The Paramecium does not have a central nervous system. (4) Amebae do not have a central nervous system.

94. **(1)** Glucagon increases blood sugar concentration by stimulating the discharge of sugar from the liver into the blood. INCORRECT CHOICES: (2) Estrogen, one of the female sex hormones, influences female sex characteristics. (3) Insulin promotes the outflow of sugar from the blood and thereby tends to decrease blood sugar concentration. (4) Testosterone, secreted by the testes, influences male secondary sex characteristics.

95. **(3)** Enzymes are specific in that they will usually only catalyze reactions involving a single substrate compound or a group of closely related compounds. INCORRECT CHOICES: (1) Salivary amylase is a protein as are all enzymes; starch is the substrate for salivary amylase. (2) Upon completion of its action, an enzyme separates from the products of the reaction and is then available to repeat its action. (4) All enzymes, including salivary amylase, are proteins.

96. **(4)** Carbon dioxide diffuses from the grasshopper's cells into the tracheae where it is transported to the external atmosphere. INCORRECT CHOICES: (1) The Malpighian tubules transport water, mineral salts, and uric acid to the digestive tube, which expels them. (2) Nephridia are the excretory tubules of earthworms. (3) Lungs are the principal means of carbon dioxide excretion in humans.

97. **(1)** Secretion of mucus provides a moist surface which permits gas exchange with the environment. INCORRECT CHOICES: (2) All organisms, regardless of habitat, produce nitrogenous waste products. (3) Enzymatic oxidation of glucose which makes available energy for cellular activity is a vital necessity of all living things. (4) A one-way digestive tract makes possible efficient processing of food by a complex, multicellular organism, regardless of habitat.

98. **(2)** In respiration, the chemical bond energy of organic molecules is transferred to adenosine triphosphate (ATP). Energy stored in ATP is available for cellular activity and is released through the action of the enzyme ATP-ase. INCORRECT CHOICES: (1) Mineral salts are inorganic compounds whose energy is not available to most living things. (3) Light energy is converted to chemical bond energy by green plants in the process of photosynthesis. (4) Nitrogenous wastes are products of metabolism containing nitrogen which must be excreted by most organisms because they are harmful.

99. **(2)** Exchange of materials including nutrients between the blood in the earthworm's closed circulatory system and its body cells is by means of diffusion through the walls of the capillaries. INCORRECT CHOICES: (1) In a closed circulatory system, the blood never leaves the blood vessels nor flows directly into the cells. (3) Nephridia are tubules which excrete water, salts, and nitrogenous wastes from the earthworm. (4) Oxygen enters the earthworm's blood through the skin but nutrients are picked up by the blood in the intestine and carried to the muscles and other tissues.

100. **(3)** The line indicating percent of undigested protein present drops in C (stomach) and D (small intestine). INCORRECT CHOICES: (1) A (mouth) and B (esophagus) secrete no protein digesting enzymes hence the percent of undigested protein stays at 100% in A and B. (2) B (esophagus) does not digest protein. (4) A (mouth) does not digest protein.

101. **(2)** C represents the stomach since, C follows A (mouth) and B (esophagus) in the digestive tract, and since the stomach contains proteases. INCORRECT CHOICES: (1) The esophagus connects the mouth and stomach; also the esophagus does not digest protein. (3) The small intestine brings about the digestion of carbohydrates and lipids as well as proteins; also, the small intestine follows the stomach. (4) Chemical digestion is largely completed by the time food reaches the large intestine.

102. **(4)** Pancreatic enzymes enter the system in D (small intestine). INCORRECT CHOICES: (1) E represents the large intestine which is not involved in the digestion of food. (2) B represents the esophagus which does not receive secretions of accessory digestive glands. (3) C represents the stomach which contains gastric glands.

103. **(4)** Capillaries in the lining of the small intestine (D) absorb digestive end products. INCORRECT CHOICES: (1) Fecal material is periodically evacuated through F (anus). (2) Most food is undigested and therefore, not capable of being absorbed, in B (esophagus). (3) Little is absorbed in C (stomach) in part because of a thick coating of mucus.

104. **(3)** E (large intestine) accumulates undigested and indigestible materials from which water is absorbed into the blood. INCORRECT CHOICES: (1) A (mouth) does not absorb water. (2) B (esophagus) does not absorb water. (4) Absorption of digestive end products into the blood occurs in D (small intestine).

105. **(1)** Food vacuoles are formed around ingested food as the Ameba engulfs the food (yeast cells). INCORRECT CHOICES: (2) Contractile vacuoles serve

in the excretion of excess water. (3) The cell wall supports and protects the plant cell. An ameba has no cell wall. (4) Buds are produced as a result of the asexual reproduction of some unicellular organisms including yeast. The Ameba does not produce buds.

106. **(1)** Respiration is the process by which the potential energy of organic molecules such as glucose is converted to the more available form of the compound ATP (adenosine triphosphate). INCORRECT CHOICES: (2) Reproduction is the ability of living organisms to produce new individuals essentially similar to themselves. (3) Digestion is the process whereby large or insoluble food materials are reduced to small, soluble molecules. (4) Excretion removes harmful or potentially harmful waste products of metabolism.

107. **(3)** The nephron is the microscopic filtering unit in a human kidney. INCORRECT CHOICES: (1) The nephridium is the filtering unit in the earthworm. (2) Malpighian tubules are the filtering units in the grasshopper. (4) The urinary bladder is a sac found in vertebrates for the storage of urine prior to its excretion.

108. **(1)** Blood in the left atrium (auricle) comes directly from the lungs where it was oxygenated. INCORRECT CHOICES: (2) Blood has left the heart and passed through blood vessels to get to the cerebrum losing some of the oxygen. (3) The nephrons of the kidneys remove nitrogenous wastes from the blood after it has passed through the body tissues. (4) The lining of the intestine is designed for absorption of digested nutrients.

109. **(4)** The rhythmic contractions of the smooth muscle fibers in the walls of the alimentary canal push the food through the digestive tract.

110. **(6)** Voluntary muscles move body parts toward and away from it by contracting and relaxing or getting shorter or longer.

111. **(1)** The rigid bones of the body are moved like levers about a fulcrum by the voluntary muscles.

112. **(3)** Ligaments are a type of connective tissue which stretch and attach bones to bones.

113. **(5)** Tendons are a type of connective tissue which does not stretch, and connects muscles to bones. INCORRECT CHOICE: (2) Cartilage tissue is found at the end of bones for protection and to ease motion of one bone on another.

114. **(2)** Skeletal muscles attached to bones permit movement of the bones resulting in locomotion in humans. INCORRECT CHOICES: (1) Setae are found on earthworms and used for locomotion. (3) Pseudopodia are formed by Amebae for locomotion; flagella are characteristic of Euglenoids. (4) Paramecia do not have muscles nor pseudopodia.

115. **(3)** During aerobic respiration, hydrogen atoms carrying energy are removed from glucose. INCORRECT CHOICES: (1), (2), and (4) Nitrogen, iron, and calcium are not present in glucose, which contains $C_6H_{12}O_6$ atoms.

116. **(2)** The cells would lose water because there would be a greater concentration of water inside the cell than outside; this would cause the cell to shrink, increasing the space between the cell membrane and cell wall. INCORRECT CHOICES: (1) Starch granules turn blue-black when treated with iodine not salt water. (3) The cells are connected by rigid cell walls and would not break down. (4) Loss of water from the cell would not cause the nucleoli to swell.

117. **(3)** In active transport energy is expended to move materials against the concentration gradient; in this case to pump water out of the cell. INCORRECT CHOICES: (1) Cyclosis is the streaming of the cytoplasm within the cell. (2) In the chemical reaction of hydrolysis, water is added in order to break down large molecules. (4) In order to take in large molecules, the cell membrane forms a pocket; this is called pinocytosis.

118. **(4)** During pinocytosis the cell membrane forms a small pocket in order to take in large molecules which cannot pass through the cell membrane. INCORRECT CHOICES: (1) Small, soluble molecules move from an area of higher to lower concentration by the process of diffusion. (2) Passive transport takes place when materials move without the expenditure of energy by the cell; this includes osmosis and diffusion. (3) Osmosis is the passive transport of water.

3

Maintenance In Plants

A. Nutrition
1. Autotrophic nutrition
   a. Photosynthesis
      (1) Process
         (a) Description
         (b) Chemical aspects
      (2) Site
   b. Chemosythesis
2. Heterotrophic nutrition

B. Digestion

C. Transport
1. Roots
   a. Root hairs
   b. Xylem
   c. Phloem
2. Stems
3. Leaves

D. Respiration
1. Types
   a. Aerobic
      (1) Description
      (2) Chemical aspects
   b. Anaerobic
      (1) Description
      (2) Chemical aspects
         (a) Alcoholic fermentation
         (b) Lactic acid fermentation
2. Energy yield

E. Excretion

F. Synthesis
   1. Process
   2. Products
      a. Structural
      b. Other
   3. Outcome

G. Regulation
   1. Production of auxins
   2. Effects of auxins

The following questions on Maintenance in Plants have appeared on previous Regents Examinations.

Directions: Each question is followed by four choices.
Underline the correct choice.

1. When do green plants carry on cellular respiration?
   (1) only during the night
   (2) only during the day
   (3) during both the night and the day
   (4) neither during the night nor during the day

2. A student placed a stalk of celery in a beaker of water which had been colored with dye. After a few hours, the student cut the stalk at the place indicated in the diagram below and observed that the dye could be seen scattered in the stem tissue at the cut location. Within which structures in the stem was the dye observed?

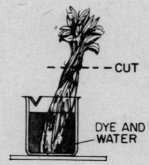

   (1) epidermal cells   (3) guard cells
   (2) phloem cells   (4) xylem cells

3. Recycled nitrogenous wastes may be used by plants to synthesize
   (1) simple sugars
   (2) glycerol
   (3) fatty acids
   (4) amino acids

4. Which type of organism can obtain energy as a result of alcoholic fermentation?
   (1) yeast
   (2) Paramecium
   (3) Hydra
   (4) earthworm

5. Chemical reactions associated with the process of aerobic respiration in green plants take place in
   (1) vacuoles
   (2) cell walls
   (3) mitochondria
   (4) ribosomes

6. The destruction of xylem tissues in a maple tree most directly interferes with the movement of
   (1) carbon dioxide out of the leaves
   (2) water to the leaves
   (3) oxygen out of the leaves
   (4) nutrients down to the roots

7. Unequal distributions of auxins in a plant bring about growth responses called
   (1) secretions
   (2) tropisms
   (3) stimuli
   (4) impulses

8. In autotrophic plants, stored starch may be converted into small organic molecules by the process of
   (1) transpiration
   (2) aerobic respiration
   (3) intracellular digestion
   (4) extracellular digestion

9. Which wavelength of light would be *least* useful to a green plant during photosynthesis?
   (1) blue
   (2) green
   (3) red
   (4) yellow

10. Within a cell of an Elodea plant, the chloroplasts move primarily as a result of
    (1) osmosis
    (2) diffusion
    (3) cyclosis
    (4) photosynthesis

11. Cells in the stem of a bean plant receive water and minerals chiefly by
    (1) diffusion through stomates
    (2) upward transport through xylem cells
    (3) upward transport through phloem cells
    (4) active transport of water through lenticels

12. Water and minerals move from the soil into a plant by the process of
    (1) diffusion, only
    (2) active transport, only
    (3) passive transport and hydrolysis
    (4) both diffusion and active transport

13. In green plants, most gas excretion takes place through
    (1) stomates and lenticels   (3) xylem and phloem cells
    (2) vacuoles and vascular tissue (4) chloroplasts and ribosomes

14. In a green plant cell, oxygen is used primarily for the process of
    (1) dehydration synthesis   (3) respiration
    (2) photosynthesis   (4) capillary action

15. The leaves on a tree normally receive their supply of carbon dioxide through the
    (1) xylem   (3) lenticels
    (2) stomates   (4) phloem

16. In a laboratory culture of yeast, it may be concluded that fermentation has occurred if chemical tests indicate the production of
    (1) carbon dioxide and water
    (2) PGAL and nitrates
    (3) oxygen and ATP
    (4) ethyl alcohol and carbon dioxide

17. When a living geranium plant is enclosed in a large jar, water droplets appear on the inner surface of the jar. These water droplets are most likely the direct result of
    (1) hydrolysis and photosynthesis
    (2) light absorption and reflection
    (3) transpiration and condensation
    (4) intracellular and extracellular digestion

18. A biologist observed a plant cell in a drop of water and illustrated it as in diagram A. He added a 10% salt solution to the slide, observed the cell, and illustrated it as in diagram B.

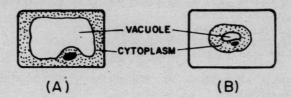

The change in appearance of the cell resulted from more
(1) salt flowing out of the cell than into the cell
(2) salt flowing into the cell than out of the cell
(3) water flowing into the cell than out of the cell
(4) water flowing out of the cell than into the cell

19. By which process are CO_2 and H_2O converted to carbohydrates?
(1) transpiration   (3) fermentation
(2) respiration   (4) photosynthesis

20. Plants excrete fewer chemical waste products than animals do. The best explanation for this is that plants
(1) do not carry on metabolic activities which produce wastes
(2) can reuse more of their waste products than animals can
(3) have no excretory mechanisms, so they store all their wastes
(4) have shorter lives than animals

21. A root hair cell may continue to absorb minerals even though the cytoplasmic concentration of these minerals is greater inside the cell than in the soil. This absorption is accomplished chiefly as a result of
(1) passive transport   (3) diffusion
(2) active transport   (4) osmosis

22. By what process does carbon dioxide pass through the stomates into the leaf?
(1) diffusion   (3) respiration
(2) osmosis   (4) pinocytosis

23. Which type of organism synthesizes organic molecules from inorganic raw materials?
(1) autotrophic   (3) saprophytic
(2) heterotrophic   (4) parasitic

24. The tissue which conducts organic food throughout a vascular plant is composed of
    (1) cambium cells
    (2) xylem cells
    (3) phloem cells
    (4) epidermal cells

25. An organism which makes its own food without the direct need for any light energy is known as a
    (1) chemosynthetic heterotroph
    (2) chemosynthetic autotroph
    (3) photosynthetic heterotroph
    (4) photosynthetic autotroph

26. During photosynthesis, molecules of oxygen are liberated from the "splitting" of water molecules. This is a direct result of the
    (1) dark reactions
    (2) light reactions
    (3) formation of PGAL
    (4) formation of CO_2

27. Which compounds include plant growth hormones?
    (1) carbohydrates
    (2) neurohumors
    (3) auxins
    (4) pigments

28. The oil of the vanilla bean and the oil of mint are flavorings found naturally in these plants. These flavorings are present in the plant cells as a result of
    (1) absorption from special mineral oils
    (2) hydrolysis of chlorophyll molecules
    (3) synthesis from simpler compounds
    (4) condensation on leaf surfaces

29. A florist who wants to speed up the formation of roots on stem cuttings will most likely apply a solution of
    (1) chlorophyll
    (2) auxins
    (3) colchicine
    (4) ribonucleic acid

30. Which organisms secrete enzymes that digest food outside their cells, and then absorb the simple end-products?
    (1) trees
    (2) grasses
    (3) fungi
    (4) algae

31. The basic raw materials for photosynthesis are
    (1) water and carbon dioxide
    (2) oxygen and water
    (3) sugar and carbon dioxide
    (4) water and oxygen

32. The primary function of root hairs in a plant is to
    (1) prevent excessive loss of water
    (2) provide increased surface area for absorption
    (3) conduct water and minerals upward
    (4) conduct organic food materials upward and downward

33. Raw materials required for autotrophic nutrition in green plants include
    (1) oxygen and glucose
    (2) oxygen and light
    (3) water and oxygen
    (4) water and carbon dioxide

34. Which would most directly be affected in a plant if root pressure and transpirational pull were not occurring?
    (1) the movement of water into the leaves
    (2) the storage of food in the roots
    (3) the passage of carbon dioxide through the epidermis
    (4) the release of energy from glucose molecules

35. Which organisms carry on heterotrophic nutrition?
    (1) ferns
    (2) grasses
    (3) fungi
    (4) mosses

36. Within a plant cell, the glucose formed as a result of photosynthesis may be used directly as
    (1) an energy source during cellular respiration
    (2) an enzyme for intracellular digestion
    (3) an absorber of radiant energy
    (4) a source of molecular oxygen

37. Bromthymol blue turns to bromthymol yellow in the presence of carbon dioxide. When the carbon dioxide is removed, the solution will return to a blue color. Two green waterplants were placed in separate test tubes, each containing water and bromthymol yellow. Both test tubes were corked. One tube was placed in the light, the other in the dark. After several days, the liquid in the tube exposed to the light turned blue.
    This demonstration illustrates that, during photosynthesis, green plants
    (1) take in carbon dioxide
    (2) need bromthymol blue
    (3) give off oxygen gas
    (4) form ATP molecules

38. Which energy conversion occurs in the process of photosynthesis?
    (1) Light energy is converted to nuclear energy.
    (2) Chemical bond energy is converted to nuclear energy.

(3) Light energy is converted to chemical bond energy.
(4) Mechanical energy is converted to light energy.

39. Plant structures that increase the surface area for absorption of water and minerals are
(1) root hairs
(2) lenticels
(3) chloroplasts
(4) guard cells

40. The raw materials used by green plants in the process of photosynthesis include carbon dioxide and
(1) pyruvic acid
(2) oxygen
(3) nitrogen
(4) water

41. Carbon dioxide, a product of plant metabolism, can be used by maple trees in the process of
(1) aerobic respiration
(2) fermentation
(3) photosynthesis
(4) transpiration

42. While looking through a microscope at a section of a leaf from a freshwater plant, a student observed some cells in which chloroplasts were moving around with the cytoplasm. This type of movement is known as
(1) pinocytosis
(2) synapsis
(3) osmosis
(4) cyclosis

43. In protists, some of the end-products of the fermentation process include
(1) ethyl alochol and lactic acid
(2) carbon dioxide and hydrogen
(3) glucose and oxygen
(4) starches and lipids

44. Which response usually occurs in plants as a result of unequal auxin distribution?
(1) Roots bend away from gravity.
(2) Stems bend toward gravity.
(3) Roots bend toward light.
(4) Stems bend toward light.

45. Which organisms are classified as heterotrophs?
(1) fungi
(2) grasses
(3) algae
(4) ferns

46. In a lake, most algae will produce oxygen as long as energy is available in the form of
(1) ADP
(2) heat
(3) light
(4) starch

47. Removing a strip of phloem from around the trunk of a tree, as shown in the diagram below, usually causes the tree to die because the roots will not receive enough

   (1) water   (3) food
   (2) minerals   (4) carbon dioxide

48. Which cells provide a continuous conducting pathway from the root of a plant to its leaves?
   (1) xylem   (3) root hairs
   (2) epidermal cells   (4) guard cells

49. During which process do root hairs use energy for the absorption of soil minerals?
   (1) simple diffusion   (3) cyclosis
   (2) active transport   (4) osmosis

50. Which is an end-product of photosynthesis?
   (1) chlorophyll   (3) light
   (2) carbon dioxide   (4) sugar

Answers

1. 3	11. 2	21. 2	31. 1	41. 3
2. 4	12. 4	22. 1	32. 2	42. 4
3. 4	13. 1	23. 1	33. 4	43. 1
4. 1	14. 3	24. 3	34. 1	44. 4
5. 3	15. 2	25. 2	35. 3	45. 1
6. 2	16. 4	26. 2	36. 1	46. 3
7. 2	17. 3	27. 3	37. 1	47. 3
8. 3	18. 4	28. 3	38. 3	48. 1
9. 2	19. 4	29. 2	39. 1	49. 2
10. 3	20. 2	30. 3	40. 4	50. 4

Explanatory Answers

1. **(3)** Respiration occurs 24 hours a day in plants as in animals. INCORRECT CHOICES: (1) Respiration converts the energy stored in chemical bonds of organic molecules into the available energy of ATP molecules. This occurs continuously in the presence of free oxygen. (2) Photosynthesis occurs only when light energy is present but respiration occurs 24 hours a day. (4) The available energy of ATP produced by respiration is required by plants both day and night.

2. **(4)** Xylem tissue conducts water and minerals upward in the plant. The dye moved upward in the xylem with water. INCORRECT CHOICES: (1) Epidermal cells form a protective layer on the outside of the stem. (2) Phloem conduct organic food materials both upward and downward. (3) Guard cells are specialized epidermal cells which make possible the exchange of gases with the external atmosphere.

3. **(4)** Amino acids contain nitrogen which in plants may come from recycled nitrogenous wastes. INCORRECT CHOICES: (1), (2), and (3) Simple sugars,

glycerol, and fatty acids do not contain nitrogen; they contain only carbon, hydrogen, and oxygen.

4. **(1)** Yeast obtains energy through a form of anaerobic respiration called fermentation; it produces CO_2 and alcohol. INCORRECT CHOICES: (2), (3), and (4) The Paramecium, Hydra, and earthworm use aerobic respiration, combining oxygen with organic compounds to release energy.

5. **(3)** Mitochondria are the sites of aerobic respiration in green plant cells as in animal cells. INCORRECT CHOICES: (1) Vacuoles are membrane-bounded spaces which act as reservoirs for water and dissolved materials. (2) Cell walls provide support and protection for plant cells. (4) Ribosomes are the sites of protein synthesis.

6. **(2)** Xylem tissue transports water and dissolved minerals upward in plants. INCORRECT CHOICES: (1) Carbon dioxide diffuses in and out through the stomates of leaves. (3) Oxygen also passes in and out through the stomates. (4) Nutrients are transported by phloem tubes.

7. **(2)** Tropisms are plant growth responses brought about by differences in auxin distribution; phototropism and geotropism are examples. INCORRECT CHOICES: (1) Secretions are substances synthesized in abundance in cells for use in metabolic activities outside the cells which produce them. (3) Stimuli are changes in the environment that bring about responses. (4) Impulses are electrochemical messages carried by nerve cells.

8. **(3)** In plants, stored nutrients are broken down by hydrolysis into smaller, usable molecules within the cell. INCORRECT CHOICES: (1) Transpiration is the loss of water through the leaves of plants. (2) Aerobic respiration is the release of energy from food not the breakdown of food molecules. (4) Extracellular digestion does not take place in autotrophic plants because they can make their own food.

9. **(2)** Chlorophyll reflects green light and therefore green light is not utilized in photosynthesis. INCORRECT CHOICES: (1) and (3) Chlorophyll traps certain wavelengths of light (red and blue light are highly effective) and converts this energy into chemical bond energy. (4) Yellow light is absorbed by chlorophyll to some extent.

10. **(3)** In cyclosis, the streaming of the cytoplasm moves the chloroplasts around in the cytoplasm. INCORRECT CHOICES: (1) Diffusion of water into and out of the cell is called osmosis. (2) Diffusion is a form of transport in which materials move from an area of higher to one of lower concen-

tration. (4) Photosynthesis is the manufacture of glucose from simple compounds using the energy of light.

11. **(2)** Xylem cells function to carry water and minerals upward. INCORRECT CHOICES: (1) Stomates do not function in transport. They are specialized for transpiration and diffusion of gases. (3) Phloem cells translocate food to roots for storage. (4) Lenticels are small corky areas on small twigs which permit the exchange of gases between the internal tissues and the environment.

12. **(4)** Water and minerals are absorbed by osmosis and diffusion, which are forms of passive transport, and by the expenditure of energy or active transport. INCORRECT CHOICES: (1) Active transport is necessary because materials must sometimes move against the gradient. (2) Water and minerals are taken into a plant by moving from areas of higher to lower concentration by osmosis or diffusion. (3) Hydrolysis is the splitting of large molecules by the addition of water.

13. **(1)** Green plants have no specialized excretory structures as do invertebrates and vertebrates. They rely on the natural processes of diffusion and osmosis. The stomates allow waste gas to pass through in solution as do the lenticels. INCORRECT CHOICES: (2) Vacuoles play many roles, such as storage; the vascular tissue is for transport. (3) Xylem and phloem cells are for transport. (4) Chloroplasts are photosynthetic; ribosomes produce proteins.

14. **(3)** In respiration, oxygen combines with organic molecules to release energy. INCORRECT CHOICES: (1) Dehydration synthesis is the chemical reaction in which molecules are joined by splitting out of water. (2) Oxygen is a product of photosynthesis. (4) Capillary action is the transport of liquids within thin tubes due to adhesion to the walls of the tubes.

15. **(2)** Stomates are openings in the leaf which permit the diffusion of carbon dioxide from the atmosphere into the leaf. INCORRECT CHOICES: (1) Xylem tissue conducts water and minerals upward in the plant. (3) Lenticels are openings in woody stems which permit gases to enter and leave the stems. (4) Phloem tissue conducts organic food materials both upward and downward.

16. **(4)** Yeast brings about the fermentation of glucose to ethyl alcohol and carbon dioxide. INCORRECT CHOICES: (1) If molecular oxygen is available, yeast may oxidize glucose to carbon dioxide and water; under these conditions fermentation may not occur. (2) PGAL is a product of pho-

tosynthesis; nitrates are used in the synthesis of amino acids. (3) The presence of oxygen makes it unlikely that fermentation will occur.

17. **(3)** The evacuation of water from the moist spongy cell layer of the leaf is necessary to move water from the root. It is also a cooling process for the internal tissues. In nature this water vapor would escape. In the bell jar, the glass sides are cooler than the escaping water, and condensation occurs. INCORRECT CHOICES: (1) Hydrolysis is digestion and photosynthesis is autotrophic nutrition. (2) Light absorption and reflection plays a major role in photosynthesis. (4) These forms of digestion are characteristic of most organisms.

18. **(4)** The plant cell in (A) contained a high concentration of water in its vacuole. When the cell was placed in a 10% salt solution, more water molecules flowed out of the cell (from a high water concentration), than into the cell (from a low, 90%, water concentration). INCORRECT CHOICES: (1) There is very little salt inside a plant cell; if all the salt left it would not bring about the shrinkage of the cell as in (B). (2) Salt flowing into the cell would cause it to swell, not to shrink. (3) Molecules such as water tend to move from high to low concentrations; furthermore, water flowing into the cell would cause it to enlarge, not to shrink.

19. **(4)** Photosynthesis is the process in which light energy is used to manufacture organic compounds from CO_2 and H_2O. INCORRECT CHOICES: (1) Transpiration is the loss of water through the stomates of leaves. (2) Respiration is the process of releasing energy from organic compounds. (3) Fermentation is a form of anaerobic respiration which produces CO_2 and water.

20. **(2)** Plants are able to use many of the products of metabolism in synthesis; they give off only those products which are no longer useful to them. INCORRECT CHOICES: (1) Plant metabolic processes produce wastes such as carbon dioxide, oxygen, water, ammonia, and others. (3) Plants have no specialized system for excretion but they reuse many waste products in synthesis; for example, carbon dioxide is used in photosynthesis. Some toxic wastes are stored in vacuoles. (4) Plants include the longest-lived organisms in the world.

21. **(2)** Active transport occurs when molecules must be moved against a concentration gradient, as here, from the low concentration in the soil to the higher cytoplasmic concentration. INCORRECT CHOICES: (1) Passive transport involves the movement of ions or molecules from a region where they occur in higher concentration to a region of lower concentration.

(3) Diffusion is a form of passive transport. (4) Osmosis is the diffusion of water into and out of cells.

22. **(1)** Carbon dioxide passes into the leaves by diffusion from an area of great concentration to one of less concentration. INCORRECT CHOICES: (2) Osmosis is the passage of water through a membrane by passive transport. (3) Respiration is the combining of oxygen with a substance which produces the release of energy. (4) Pinocytosis is the folding in of a cell's surface for the absorption of large organic molecules.

23. **(1)** An autotrophic organism meets it nutritional needs by manufacturing its own organic molecules from inorganic raw materials, usually by means of photosynthesis. INCORRECT CHOICES: (2) Heterotrophic organisms must consume preformed organic compounds. (3) Saprophytic organisms are heterotrophic plants which live on dead organic matter. (4) Parasites are heterotrophs which live in close association with other organisms, their hosts; parasites benefit at the expense of the hosts.

24. **(3)** Phloem cells form the tubes which conduct dissolved organic food from the leaves through the stems and roots in plants. INCORRECT CHOICES: (1) Cambium cells are the growth cells located in the stems beneath the epidermis. (2) Xylem cells form tubes which conduct water through a vascular plant—from roots through stems and leaves. (4) Epidermal cells protect the inner structures of a plant and absorb water in the roots.

25. **(2)** A chemosynthetic autotroph is an organism which forms needed organic food compounds from the use of chemical energy. INCORRECT CHOICES: (1) and (3) Heterotrophs cannot utilize either chemical or light energy to make nutrients. (4) A photosynthetic autotroph produces food by directly using light energy.

26. **(2)** During the light phase of photosynthesis, water is split into hydrogen and oxygen molecules. INCORRECT CHOICES: (1) In the dark phase, carbon dioxide and hydrogen are combined forming organic compounds. (3) PGAL is an intermediate compound formed during the dark reaction. (4) Carbon dioxide is formed during respiration.

27. **(3)** Auxins function as plant growth regulators. INCORRECT CHOICES: (1) Carbohydrates are a large group of organic compounds including glucose, sucrose, starch, and cellulose. (2) Neurohumors are secreted by nerve cells and make possible the passage of the impulse across a synapse. (4) Pigments such as chlorophyll absorb specific wavelengths of light and therefore, appear colored.

28. **(3)** Plants synthesize many substances including flavorings. Lipids such as the oils of the vanilla bean and mint are made from simpler compounds such as fatty acids and glycerol. INCORRECT CHOICES: (1) Plants absorb water and minerals from the soil but not organic compounds such as oils. (2) The hydrolysis or breakdown of chlorophyll does not yield flavorings such as the oils of vanilla and mint. (4) The oil of the vanilla bean and the oil of mint are synthesized within these plants and do not condense (change from gas to liquid) from the atmosphere.

29. **(2)** Auxins are plant hormones which promote growth. INCORRECT CHOICES: (1) Chlorophyll enables a plant to capture light in order to make food. (3) Colchicine is a mutagenic agent which produces plants with extra sets of chromosomes (polyploidy). (4) Ribonucleic acid or RNA is found in all plant and animal cells.

30. **(3)** Fungi are heterotrophic plants which lack chlorophyll. They obtain preformed organic molecules from other organisms by extracellular digestion and absorption. INCORRECT CHOICES: (1), (2), and (4) Trees, grasses, and algae are autotrophic organisms containing chlorophyll and can synthesize their own food molecules.

31. **(1)** Water + carbon dioxide $\xrightarrow[\text{chlorophyll enzymes}]{\text{energy}}$ glucose + oxygen. INCORRECT CHOICES: (2) Oxygen and water are end products of photosynthesis. (3) Sugar (glucose) is an end product of photosynthesis. (4) Water and oxygen are end products of photosynthesis.

32. **(2)** Root hairs are epidermal tissue and as such are extremely thin. This adaptation allows them to become the major water absorbers of the root. The main root is covered in a cork cambium layer which does not lend well to water absorption. INCORRECT CHOICES: (1) Through transpirational pull, water would not be lost from the roof surface. (3) This function is the job of the xylem. (4) This function is the job of the phloem.

33. **(4)** Photosynthesis, or autotrophic nutrition in green plants, uses water and carbon dioxide as raw materials. INCORRECT CHOICES: (1) Oxygen and glucose are the products of photosynthesis or autotrophic nutrition in plants. (2) Light but not oxygen is required for photosynthesis. (3) Water is required for photosynthesis, oxygen is not directly required.

34. **(1)** Two of the main upward forces which get water to the leaves are root pressure and transpirational pull. INCORRECT CHOICES: (2) Food gets to the roots by diffusion and gravitational force. (3) Carbon dioxide enters the

plant from the atmosphere. (4) The release of energy from glucose is a chemical change not a physical one.

35. **(3)** Fungi cannot make their own food. They digest organic substances outside their own structures. INCORRECT CHOICES: (1), (2), (3) Ferns, grasses, and mosses contain chlorophyll and make their own food autotrophically during photosynthesis.

36. **(1)** Glucose, through aerobic respiration, is used to liberate, through oxidation, ATP molecules which in turn are used for energy. A green plant manufactures its own glucose. INCORRECT CHOICES: (2) Glucose cannot be used as an enzyme in that it is not protein in nature nor can it catalyze any reaction. (3) The absorbers of radiant energy which are most efficient are the chlorophyll molecules of green plant cells. (4) If glucose were dehydrated, it would liberate water molecules. If glucose were oxidized it would liberate CO_2 and form H_2O and ATP molecules. No molecular oxygen is formed.

37. **(1)** Since bromothymol blue turns yellow in the presence of CO_2, then placing a green plant in bromothymol yellow and sunlight would necessarily find the following: The plant absorbs the CO_2 from solution and as this occurs, the solution turns blue. INCORRECT CHOICES: (2) Bromothymol blue was not placed with the plants. The only variable in the experiment is the carbon dioxide. (3) The question does not say anything about oxygen. It is formed as a by-product in photosynthesis. (4) ATP molecules are formed in respiration.

38. **(3)** Plants utilize light energy for photosynthesis to form chemical bonds in making glucose. INCORRECT CHOICES: (1) Light energy is not converted to nuclear energy during photosynthesis. (2) Chemical bond energy is not converted to nuclear energy during photosynthesis. (4) Mechanical energy is not converted to light energy during photosynthesis.

39. **(1)** Root hairs are elongated epidermal cells which increase the surface area of the root. INCORRECT CHOICES: (2) Lenticels are openings in woody stems which permit diffusion of gases. (3) Chloroplasts are organelles in which photosynthesis occurs. (4) Guard cells are specialized epidermal cells which control the size of stomates in the leaf.

40. **(4)** Carbon dioxide and water are used in the chloroplasts to form simple sugars and oxygen. INCORRECT CHOICES: (1) Pyruvic acid is a product of the anaerobic phase of the respiration of glucose. (2) Oxygen is a by-product of photosynthesis. (3) Nitrogen is not used in photosynthesis.

41. **(3)** Carbon dioxide is one of the raw materials for photosynthesis. IN-

CORRECT CHOICES: (1) Aerobic respiration utilizes glucose as a raw material and provides carbon dioxide as a product. (2) Fermentation uses glucose as a raw material. (4) Transpiration is the removal of the water molecules through the stomates, and is a purely physical process.

42. **(4)** Cyclosis is the movement of materials and structures within a cell. INCORRECT CHOICES: (1) Pinocytosis is the process of taking in liquid and small molecules by the cells. (2) Synapsis is the lining up of homologous chromosomes in pairs during the first meiotic division. (3) Osmosis is the passage of water in and out of the cell.

43. **(1)** Ethyl alcohol and carbon dioxide or lactic acid are produced as a result of anaerobic respiration in protists. INCORRECT CHOICES: (2) Carbon dioxide, but not hydrogen, is a product of fermentation or anaerobic respiration. (3) Glucose and oxygen are the raw materials for aerobic respiration. (4) Starches and lipids are organic compounds made to store energy.

44. **(4)** Auxins are destroyed on the side of the stem facing the light; this unequal distribution causes the stem to bend toward the light. INCORRECT CHOICES: (1) and (2) Plant hormones cause roots to bend toward gravity and stems to bend away from gravity; this is called geotropism. (3) Roots are not affected by light.

45. **(1)** Fungi are nongreen plants such as mushrooms. They cannot carry on photosynthesis and must take in preformed organic compounds and are, therefore, classified as heterotrophs. INCORRECT CHOICES: (2), (3), and (4) Grasses, algae, and ferns all contain chlorophyll, carry on photosynthesis, and are autotrophs.

46. **(3)** Algae are one-celled green plants; they require light energy to produce oxygen by the process of photosynthesis. INCORRECT CHOICES: (1) ADP combines with phosphate to store energy in the form of ATP. (2) Heat energy cannot be used for photosynthesis. (4) Starch is a polysaccharide made by plants to store food.

47. **(3)** Food or dissolved sugar are transported by the phloem tubes located under the bark. Removing a strip of phloem interrupts the movement of stored sugar in the tree. INCORRECT CHOICES: (1) and (2) Water and minerals are transported by the xylem tubes located toward the center of the tree. (4) Carbon dioxide enters and leaves through openings in the leaf called stomates.

48. **(1)** Xylem tissue transports water and dissolved minerals upward from the roots to the leaves. INCORRECT CHOICES: (2) Epidermal cells are protective

living cells found on the outside of organs such as roots. (3) Root hairs are microscopic projections which increase the root's surface area for absorption. (4) Guard cells control the opening and closing of the stomates in the leaves.

49. **(2)** Active transport requires the expenditure of energy in order to move molecules from an area of lower to higher concentration. INCORRECT CHOICES: (1) Simple diffusion does not require the expenditure of energy; molecules move from an area of higher to lower concentration by molecular motion. (3) Cyclosis is the streaming of the cytoplasm and does require energy; it is not involved in the absorption of minerals. (4) Osmosis is the passive transport of water and does not require the root hair to use energy.

50. **(4)** Glucose, a simple sugar, is a product of photosynthesis. INCORRECT CHOICES: (1) Chlorophyll is needed to trap light energy for photosynthesis. (3) Light supplies the energy needed. (2) Carbon dioxide is a raw material of photosynthesis.

4

Reproduction and Development

A. Cell Division
   1. Mitotic cell division
      a. Processes
         (1) Mitosis
         (2) Cytoplasmic division
      b. Comparison between plant and animal cell division
   2. Meiotic cell division
      a. Process
      b. Comparison with mitotic cell division

B. Reproduction
   1. Asexual
      a. Types
      b. Results
   2. Sexual
      a. Animals
         (1) Gamete formation
            (a) Spermatogenesis
            (b) Oogenesis
         (2) Zygote formation
            (a) External fertilization
            (b) Internal fertilization
      b. Plants
         (1) Gamete formation
            (a) Spermatogenesis
            (b) Oogenesis
         (2) Zygote formation
            (a) Pollination
            (b) Fertilization
         (3) Embryo formation

C. Development
  1. Animals
    a. Embryonic development
      (1) Cleavage and differentiation
      (2) Site of embryonic development
        (a) External: in water, on land
        (b) Internal: placental animals, nonplacental animals
    b. Postembryonic development
  2. Plants
    a. Seed formation
    b. Seed development
D. Reproduction and Development in Man
  1. Reproduction
    a. Gamete formation
      (1) Male reproductive system
        (a) Structure
        (b) Function
      (2) Female reproductive system
        (a) Structure
        (b) Function: hormones, menstrual cycle
    b. Zygote formation
  2. Development
    a. Prenatal
    b. Postnatal

The following questions on Reproduction and Development have appeared on previous Regents Examinations.

Directions: Each question is followed by four choices.
Underline the correct choice.

1. External fertilization and external development are characteristic of most
   (1) fishes   (3) reptiles
   (2) birds   (4) mammals

2. In plants, which tissue is best described as the lateral meristem region?
   (1) phloem   (3) xylem
   (2) cambium   (4) pith

3. In which structure of a plant would a developing seed be found?
   (1) ovary   (3) stigma
   (2) anther   (4) pollen

4. Unfertilized eggs of a frog can be made to undergo cleavage if the eggs are pricked with a needle. This type of development is known as
   (1) parthenogenesis
   (2) metamorphosis
   (3) differentiation
   (4) pinocytosis

5. Which plant structure serves a function most similar to the production of sperm by a testis?
   (1) stamen
   (2) stigma
   (3) ovule
   (4) pistil

6. Stimulation of a follicle in the human female involves a hormone secreted by the
   (1) pituitary gland
   (2) adrenal gland
   (3) uterus
   (4) ovary

7. Which is the correct order of processes in the formation of an animal embryo?
   (1) fertilization → cleavage → gastrulation → differentiation
   (2) cleavage → fertilization → gastrulation → differentiation
   (3) fertilization → differentiation → cleavage → gastrulation
   (4) cleavage → gastrulation → fertilization → differentiation

Directions: Base your answers to questions 8 and 9 on the diagram below which represents one of the early stages in the development of an embryo.

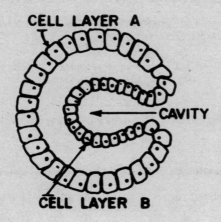

CELL LAYER A

CAVITY

CELL LAYER B

8. By which process was the cavity formed?
   (1) fertilization   (3) gastrulation
   (2) gestation   (4) ovulation

9. Which tissues in the human embryo will be formed as a result of differentiation of cell layer *B*?
   (1) nerves and brain   (3) epidermis of the skin
   (2) sense organs   (4) lining of the stomach

10. What is one advantage of sexual reproduction over asexual reproduction to a species?
    (1) production of more offspring
    (2) production of larger offspring
    (3) faster development of offspring
    (4) greater variation among offspring

11. In humans, a single primary sex cell may produce four gametes. These gametes are known as
    (1) diploid egg cells   (3) polar bodies
    (2) monoploid egg cells   (4) sperm cells

12. If an animal has a placenta, it will also have
    (1) an exoskeleton   (3) a large yolk sac
    (2) a shell   (4) an umbilical cord

13. This diagram represents a stage of mitotic cell division. What is the diploid number of chromosomes in the cell before cell division?

    (1) 5   (3) 8
    (2) 2   (4) 4

14. Artificial parthenogenesis may be accomplished by stimulating
    (1) an ovary   (3) a spore
    (2) an unfertilized egg   (4) a pollen grain

15. Which is an important adaptation for reproduction among land animals?
    (1) fertilization of gametes outside the body of the female
    (2) fertilization of gametes within the body of the female
    (3) production of sperm cells with thick cell walls
    (4) production of sperm cells with thin cell walls

Directions: Base your answers to questions 16 through 20 on your knowledge of biology and the diagrams below which represent four stages of the reproductive process.

16. Which process immediately follows stage 4?
    (1) fertilization   (3) gastrulation
    (2) cleavage   (4) metamorphosis

17. In which stage(s) are the nuclei diploid?
    (1) 1, only   (3) 2 and 3, only
    (2) 1 and 2, only   (4) 2, 3, and 4

18. Which stage is the direct result of *many* mitotic cell divisions?
    (1) 1   (3) 3
    (2) 2   (4) 4

19. During which stage is the organism called a zygote?
    (1) 1   (3) 3
    (2) 2   (4) 4

20. Which stage is a direct result of meiotic cell division?
    (1) 1   (3) 3
    (2) 2   (4) 4

21. A plant cell with 12 chromosomes undergoes normal mitosis. What is the total number of chromosomes in each of the resulting daughter cells?
    (1) 24   (3) 6
    (2) 12   (4) 4

Directions (22–24): For each statement in questions 22 through 24, select the biological structure, chosen from the list below, that is best described by that statement.

Biological Structures

(1) Allantois
(2) Amnion
(3) Uterus
(4) Yolk sac

22. This structure contains fluid which provides a watery environment and protection from shock for a developing embryo.

23. This structure serves as the major organ of excretion of nitrogenous wastes for the chicken embryo.

24. This structure provides food for a bird embryo.

25. The transfer of structures containing monoploid nuclei from the anther of the stamen to the stigma of the pistil is called
    (1) fertilization
    (2) transpiration
    (3) crossing-over
    (4) pollination

26. In frogs, zygote formation normally occurs in the
    (1) ovary
    (2) oviduct
    (3) external environment
    (4) internal environment

27. An organism having 48 chromosomes in its body cells will form gametes which have a total of
    (1) 12 homologous pairs of chromosomes
    (2) 24 homologous pairs of chromosomes
    (3) 12 chromosomes
    (4) 24 chromosomes

28. Which structure in a bean seed develops into the leaves and upper part of the stem?
    (1) ovule
    (2) epicotyl
    (3) hypocotyl
    (4) cotyledon

29. Asexual reproduction *differs* from sexual reproduction in that, in asexual reproduction,
    (1) new organisms are usually genetically identical to the parent
    (2) the reproductive cycle involves the production of gametes
    (3) nuclei of sex cells fuse to form a zygote
    (4) offspring show much genetic variation

30. In hermaphrodites, meiosis occurs in the
    (1) ovaries, only
    (2) testes, only
    (3) ovaries and testes
    (4) uterus and ovaries

31. Which structure provides food for the developing plant embryo?
    (1) cotyledon
    (2) seed coat
    (3) hypocotyl
    (4) pollen grain

32. In cells produced as a result of meiosis, the chromosome number
    (1) is reduced to half the original number
    (2) remains the same as the original number
    (3) is restored to the original number
    (4) is twice the original number

33. From which structure in the seed does the plant root develop?
    (1) epicotyl
    (2) cotyledon
    (3) ovule
    (4) hypocotyl

34. Asexual reproduction is most closely associated with
    (1) mitotic cell division
    (2) fusion of nuclei
    (3) formation of gametes
    (4) reduction of chromosome number

35. The pairing of homologous chromosomes during meiosis is called
    (1) alignment
    (2) disjunction
    (3) synapsis
    (4) crossing-over

36. Which statement describes the process of budding in a unicellular organism?
    (1) Fertilization is preceded by meiosis.
    (2) There is an equal division of the cytoplasm.
    (3) The cytoplasm divides unequally.
    (4) Fertilization is preceded by mitosis.

37. In humans, gestation normally occurs while a developing organism is in the
    (1) uterus
    (2) follicle
    (3) vagina
    (4) ovary

38. In sexual reproduction, the 2n chromosome number is restored as a direct result of
    (1) fertilization
    (2) gamete formation
    (3) cleavage
    (4) meiosis

39. During the development of a chicken egg, which embryonic membrane is used for both respiration and excretion?
    (1) amnion
    (2) allantois
    (3) placenta
    (4) yolk sac

40. In which structure of a geranium plant does fertilization occur?
    (1) stem
    (2) stamen
    (3) ovule
    (4) anther

41. In which structure does a pollen grain form?
    (1) ovary
    (2) anther
    (3) stigma
    (4) root

42. At fertilization, the normal diploid number of chromosomes is restored by the fusion of the nuclei of
    (1) zygotes, only
    (2) gametes, only
    (3) body cells, only
    (4) zygotes, gametes, and body cells

43. Mitotic and meiotic cell division are similar in that both processes
    (1) produce diploid gametes from monoploid cells
    (2) produce monoploid gametes from diploid cells
    (3) involve synapsis of homologous chromosomes
    (4) involve replication of chromosomes

44. Which structure of a vascular plant contains monoploid nuclei?
    (1) leaf bud
    (2) ovule
    (3) root hair
    (4) seed

45. Flowers whose reproductive structures consist only of stamens would be able to produce
    (1) fruits with seeds
    (2) fruits without seeds
    (3) pollen
    (4) ovules

46. After normal mitotic division, how many chromosomes does each new daughter cell contain as compared to the mother cell?
    (1) the same number
    (2) twice as many
    (3) half as many
    (4) four times as many

47. Which represents the proper sequence in the development of a flowering plant?
    (1) germination → pollination → fruit formation → fertilization
    (2) fertilization → gastrula → blastula → embryo
    (3) fertilization → zygote → gastrula → embryo → fetus
    (4) pollination → fertilization → seed formation → seed germination

Directions: Base your answers to questions 48 and 49 on the diagram below which represents the reproductive cycle of animals.

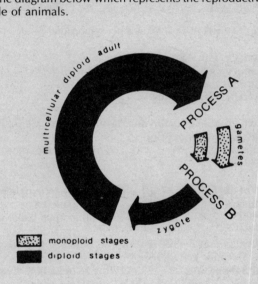

monoploid stages
diploid stages

48. Where does process A occur in the human female?
    (1) ovary   (3) placenta
    (2) uterus   (4) vagina

49. In the life cycle of a frog, process B takes place in the
    (2) ovary   (3) water
    (2) oviduct   (4) soil

50. Which structure of a seed provides the developing plant embryo with food?
    (1) pollen grain   (3) hypocotyl
    (2) epicotyl   (4) cotyledon

51. Cell A and cell B are both in the final stages of cell division. Which one represents budding?

CELL A   CELL B

(1) cell A, only   (3) both cell A and cell B
(2) cell B, only   (4) neither cell A nor cell B

52. The earthworm is classified as a hermaphrodite because it has the ability to
    (1) produce both eggs and sperm
    (2) produce eggs that develop without fertilization
    (3) reproduce asexually
    (4) reproduce by budding

53. From which human embryonic layer do the epidermis of the skin and the nervous system develop?
    (1) ectoderm   (3) mesoderm
    (2) blastula   (4) gastrula

54. Which of the following least affects the human female menstrual cycle?
    (1) pituitary   (3) pancreas
    (2) ovary   (4) corpus luteum

55. As a fertilized egg develops into an embryo, it undergoes a series of mitotic divisions known as
    (1) parthenogenesis   (3) reduction division
    (2) meiosis   (4) cleavage

56. A woman gave birth to triplets, two identical girls and one boy. The number of egg cells involved would be
    (1) 1   (3) 3
    (2) 2   (4) 4

57. A certain fruit tree is found to have desirable characteristics. These characteristics could be propagated most easily and quickly by
    (1) planting seeds produced by the tree
    (2) grafting cuttings taken from the tree
    (3) cross-pollinating with another tree
    (4) culturing highly differentiated cells of the tree

58. Which type of tissue is found in both the cambium and the growing tip of the root of a plant?
    (1) epidermal
    (2) meristematic
    (3) phloem
    (4) xylem

59. In seed plants, an increase in the number of cells is restricted largely to specific regions known as
    (1) stomates
    (2) amnions
    (3) endoderms
    (4) meristems

60. Which membrane is both a protective sac and a container for the fluid in which an embryo is suspended?
    (1) chorion
    (2) placenta
    (3) allantois
    (4) amnion

Directions: Base your answers to questions 61 through 62 on the diagram of the bean seed below.

61. Which structure develops into the upper part of the stem in the adult plant?
    (1) 1
    (1) 2
    (3) 3
    (4) 4

62. Which structure provides food for the developing embryo?
    (1) 1
    (2) 2
    (3) 3
    (4) 4

63. Which process reduces the chromosome number from diploid (2n) to monoploid (n)?
    (1) meiosis
    (2) mitosis
    (3) polyploidy
    (4) fertilization

64. In animals, the process which results in monoploid gametes is known as
    (1) meiosis
    (2) mitosis
    (3) fertilization
    (4) fission

65. An earthworm is an example of a hermaphrodite because it
    (1) develops by metamorphosis
    (2) inhabits a terrestrial environment
    (3) has functional male and female reproductive organs
    (4) carries on internal fertilization and development

Directions: Base your answers to questions 66 and 67 on the diagram below of a developing chicken embryo.

66. Where would the bulk of the stored food be located?
    (1) 5   (3) 3
    (2) 6   (4) 4

67. In which area would amniotic fluid be found?
    (1) 1   (3) 6
    (2) 2   (4) 4

Directions: Base your answers to questions 68 and 69 on the diagrams below.

(A)   **(B)**   **(C)**   **(D)**

68. Which diagram shows the formation of a cell plate?
    (1) *A*   (3) *C*
    (2) *B*   (4) *D*

69. Which diagram shows the pairing of homologous chromosomes?
    (1) A   (3) C
    (2) B   (4) D

70. A single cell reproduces as shown in the diagram below. Which type of reproduction is shown?

PARENT
CELL

    (1) spore formation   (3) fertilization
    (2) budding   (4) seed formation

71. A male fish produces gametes called
    (1) egg cells   (3) testes
    (2) sperm cells   (4) zygotes

72. Which condition is *not* necessary for the germination of a seed?
    (1) sufficient oxygen   (3) proper soil
    (2) sufficient moisture   (4) proper temperature

73. In the Paramecium, which is true of a daughter cell that results from fission?
    (1) It has one-half as many chromosomes as the parent cell.
    (2) It has twice as many chromosomes as the parent cell.
    (3) It is the same size as the parent cell, but has fewer chromosomes than the parent cell.
    (4) It is smaller than the parent cell, but contains the same number of chromosomes as the parent cell.

74. Homologous pairs of chromosomes are *not* normally found in
    (1) zygotes   (3) gametes
    (2) body tissue cells   (4) embryonic nerve cells

75. In an animal cell undergoing mitotic cell division, which of the following events does *not* normally occur?
    (1) cell plate formation   (3) centromere replication
    (2) spindle fiber production   (4) chromosomal disjunction

76. Inside the ovule of a plant, the zygote undergoes a series of mitotic divisions which result in a multicellular
    (1) fruit   (3) spore
    (2) embryo   (4) seed

77. Structures that hold chromatids together in double-stranded chromosomes are known as
    (1) centrioles
    (2) polar bodies
    (3) centromeres
    (4) spindle fibers

78. In flowering plants, sperm nuclei are formed in
    (1) testes
    (2) ovules
    (3) pollen tubes
    (4) polar bodies

79. In plant reproduction, the union of the egg nucleus with the sperm nucleus is known as
    (1) pollination
    (2) parthenogenesis
    (3) fertilization
    (4) cleavage

80. A plant structure which normally contains the monoploid number of chromosomes is the
    (1) root hair
    (2) phloem cell
    (3) embryo
    (4) pollen grain

81. Which process occurs in sexual reproduction?
    (1) fission
    (2) budding
    (3) vegetative propagation
    (4) fusion of gametes

82. Which form of reproduction could produce five different kinds of apples on one apple tree?
    (1) bulb formation
    (2) fission
    (3) spore formation
    (4) grafting

Answers

1. 1	18. 4	35. 3	52. 1	69. 3
2. 2	19. 2	36. 3	53. 1	70. 2
3. 1	20. 1	37. 1	54. 3	71. 2
4. 1	21. 2	38. 1	55. 4	72. 3
5. 1	22. 2	39. 2	56. 2	73. 4
6. 1	23. 1	40. 3	57. 2	74. 3
7. 1	24. 4	41. 2	58. 2	75. 1
8. 3	25. 4	42. 2	59. 4	76. 2
9. 4	26. 3	43. 4	60. 4	77. 3
10. 4	27. 4	44. 2	61. 1	78. 3
11. 4	28. 2	45. 3	62. 3	79. 3
12. 4	29. 1	46. 1	63. 1	80. 4
13. 2	30. 3	47. 4	64. 1	81. 4
14. 2	31. 1	48. 1	65. 3	82. 4
15. 2	32. 1	49. 3	66. 2	
16. 3	33. 4	50. 4	67. 4	
17. 4	34. 1	51. 1	68. 4	

Explanatory Answers

1. **(1)** Animals that live in the water are characterized by external fertilization and external development. INCORRECT CHOICES: (2) and (3) Birds and reptiles have internal fertilization followed by external development. (4) In mammals, both fertilization and development take place internally.

2. **(2)** The cambium layer provides for the expansion of the tissue by mitotic growth. The lateral (outward, or growth in girth) cambium would add width to trees. INCORRECT CHOICES: (1) The phloem is a transport tissue which relays food produced in the leaves to the roots. (3) The xylem is

a transport tissue also which relays water and minerals to the leaf. (4) The pith is a storage tissue of roots and stems consisting of parenchyma cells.

3. (1) A seed is a ripened ovule containing an embryo and as such would be found in the ovary. INCORRECT CHOICES: (2) The anther is the male reproductive organ. (3) The stigma of the pistil traps for the pollen grains. (4) The pollen are the male gametes.

4. (1) Parthenogenesis is the embryonic development of an egg which has not been fertilized by a sperm cell. INCORRECT CHOICES: (2) Metamorphosis is a process of change in form or appearance of an organism as it progresses from the egg to the adult stage. (3) Differentiation is the process whereby changes in embryonic cells occur to form the various tissues and organs of an offspring. (4) Pinocytosis is the folding-in of a cell surface to form a sac for the absorption of large organic molecules.

5. (1) The stamen is the male reproductive structure which produces pollen or the male gamete. INCORRECT CHOICES: (2) The stigma is the site of pollination. (3) The ovule is a structure in the ovary from which the seed will develop. (4) The pistil is the female reproductive part of the flower.

6. (1) The pituitary gland secretes follicle stimulating hormone (FSH). Its function is to stimulate the ovary to produce an egg through stimulation of the follicle. INCORRECT CHOICES: (2) The adrenal gland produces a hormone called adrenalin which raises the amount of insulin in the body. (3) The uterus does not secrete any hormones but it is the site for the developing embryo. (4) The ovary produces estrogen which stimulates the uterus into developing its primary lining, and eventually the ovary produces another hormone, progesterone, which further causes the uterus to develop.

7. (1) The process of formation of animal embryos must begin with fertilization. The only two choices are 1 and 3. Of those two, only number 1 is in correct order. INCORRECT CHOICES: (2) and (4) Since fertilization does not begin these two choices we have to eliminate them. (3) Even though fertilization begins this sequence, cleavage and differentiation are reversed in order.

8. (3) The cavity was obviously formed by a pushing-in force. This force is gastrulation. INCORRECT CHOICES: (1) Fertilization is the union of sperm and egg. (2) Gestation is the length of time the mother nourishes and protects the young in the uterus. (4) Ovulation is the release of an egg from the follicle.

9. **(4)** Cell layer *B* is pointing to the endoderm which does produce the lining of the stomach. INCORRECT CHOICES: (1) The nerves and brain are produced by the ectoderm which is cell layer *A*. (2) The sense organs are produced by the ectoderm in that they are also part of the nervous system. (3) The epidermis is produced by the ectoderm also.

10. **(4)** Asexual reproduction provides for the exact duplication of chromosomes in the offspring that were in the parent cell. Sexual reproduction, through meiosis and fertilization, provides for variety in the offspring and thus allows the offspring a better chance to adapt. INCORRECT CHOICES: (1) Neither method allows for the greater production of offspring than the other. (2) Again, neither method competes in size requirements. (3) Faster development among asexually reproducing organisms is not an advantage over slower development.

11. **(4)** Four secondary sex cells called sperm cells are formed from one primary sex cell in the testis. INCORRECT CHOICES: (1) An egg cell is not diploid. (2) Only one monoploid egg cell is formed from a primary sex cell of the ovary. (3) Polar bodies are not gametes.

12. **(4)** The placenta is a mammalian characteristic. The only other choice listed which is a mammalian characteristic is the umbilical cord. INCORRECT CHOICES: (1), (2), and (3) All of these choices refer to nonmammalian structures.

13. **(2)** The diagram shows the movement of two single-stranded chromosomes to opposite ends of the spindle. INCORRECT CHOICES: (1) and (3) Only four chromosomes are shown in the cell. (4) Of the four single-stranded chromosomes shown, only two are moving to each end of the spindle; the diploid number is the same before and after mitotic cell division.

14. **(2)** Parthenogenesis is the development of an unfertilized egg. INCORRECT CHOICES: (1) An ovary is a female reproductive organ which produces the female gametes, eggs. (3) Spores are single cells which can develop into new individuals. (4) Pollen grains are produced by the anther, part of the male reproductive organ of a flower.

15. **(2)** Internal fertilization usually increases the chances for fertilization and the formation of offspring. INCORRECT CHOICES: (1) It is less likely that the gametes would meet to fuse in a nonconfined space. (3) Thick-walled sperms would not as easily fuse with the ova. (4) Thin-walled sperms would most likely fracture prior to fusing with the ova.

16. **(3)** Stage 4 represents the blastula, one side of which becomes indented during gastrulation. INCORRECT CHOICES: (1) Fertilization, the union of egg

and sperm, occurs between stages 1 and 2. (2) Cleavage produces the blastula, stage 4. (4) Metamorphosis, the change from larva to adult, is not represented here.

17. **(4)** Fertilization restores the diploid number, and the mitotic cell divisions which follow fertilization maintain the diploid number in all subsequent stages. INCORRECT CHOICES: (1) In stage 1, the nuclei of sperm and egg are monoploid as a result of meiosis. (2) The nuclei of stage 1 are monoploid while the nucleus of stage 2 is diploid. (3) The nuclei of stages 2 and 3 are diploid but so are the nuclei of stage 4.

18. **(4)** Stage 4, the blastula, is the result of many mitotic cell divisions known as cleavage. INCORRECT CHOICES: (1) The sperm and egg of stage 1 are produced by meiotic cell division. (2) Stage 2 is formed by the union of egg and sperm. (3) One mitotic cell division produces stage 3 from the zygote, stage 2.

19. **(2)** The zygote, stage 2, is formed by the union of egg and sperm. INCORRECT CHOICES: (1) Stage 1 represents the sperm and egg. (3) Stage 3 represents a two-cell embryo. (4) Stage 4 represents a blastula.

20. **(1)** The sperm and egg of stage 1 are produced by meiotic cell division. INCORRECT CHOICES: (2) Stage 2 is produced by fertilization. (3) Stage 3 is produced by mitotic cell division. (4) Stage 4 is produced by many mitotic cell divisions known as cleavage.

21. **(2)** In mitosis, the diploid number of chromosomes is maintained in the cells formed. INCORRECT CHOICES: (1) This is double the diploid number. (3) This is half the diploid number which would be found in the secondary sex cells. (4) This is one-third the diploid number.

22. **(2)** The amnion is an embryonic membrane containing fluid which provides a watery environment and protection from shock. INCORRECT CHOICES: (1) The allantois functions in respiration and excretion. (3) The uterus serves as a place where the embryo develops in placental mammals. (4) Blood vessels in the yolk sac transport food from the yolk to the chick embryo.

23. **(1)** The allantois accumulates nitrogenous wastes (uric acid) during the development of the chick embryo. INCORRECT CHOICES: (2) The amnion provides a watery environment and protection from shock. (3) and (4) See number 22 above.

24. **(4)** See answer 22 above. INCORRECT CHOICES: (1), (2), and (3) See number 22 above.

25. **(4)** Pollination refers to the transfer of pollen grains (containing monoploid nuclei) from the anther to the stigma. INCORRECT CHOICES: (1) Fertilization refers to the union of male and female nuclei; in flowers this occurs within the ovule. (2) Transpiration refers to the loss of water from the leaves of a plant. (3) Crossing-over involves the exchange of portions of one chromatid for corresponding portions of another during meiosis.

26. **(3)** Zygote formation or fertilization in frogs normally occurs in the water of a pond or stream. INCORRECT CHOICES: (1) The ovary produces monoploid egg cells. (2) The frog oviduct secretes a jelly coating around the unfertilized eggs. (4) Fertilization in frogs is external, not internal.

27. **(4)** Gametogenesis is the reduction division (meiosis) to form gametes. Therefore the number 48 represents the diploid number and 24 represents the monoploid number. INCORRECT CHOICES: (1) The number 12 represents one-fourth the diploid number and also the word homologous appears, which means there are a pair of fundamentally similar chromosomes present. We find in gamete production that only one of the homologues is present. (2) The number 24 is correct but again the word homologous appears. (3) Twelve single chromosomes would represent again one-fourth the prescribed number.

28. **(2)** The epicotyl develops into the leaves and the upper part of the stem. INCORRECT CHOICES: (1) The ovule develops into a seed containing a plant embryo. (3) The hypocotyl develops into the root and the lower part of the stem. (4) The cotyledon stores food for the developing plant embryo.

29. **(1)** The genetic traits would be identical because only one set of genes from one sex cell is inherited by the organisms. INCORRECT CHOICES: (2) Asexual reproduction may involve somatic or body cells. (3) Sexual reproduction results in the formation of a zygote. (4) Very little genetic variation is apparent because only one cell is used.

30. **(3)** Hermaphrodites are organisms with both functional male and female gonads (testes and ovaries). INCORRECT CHOICES: (1) and (2) In hermaphrodites, both ovaries and testes produce gametes. (4) Placental female mammals have ovaries which produce eggs and the uterus in which internal development takes place.

31. **(1)** The cotyledon in the seed stores food for the developing plant embryo. INCORRECT CHOICES: (2) The seed coat protects the embryo in the seed. (3) The hypocotyl develops into the root and lower portion of the stem. (4) A pollen grain forms sperm nuclei, one of which will unite with the egg nucleus in the ovule to form a zygote.

32. **(1)** As a result of meiosis, the diploid chromosome number is reduced to half the original number, hence, the cells produced have the monoploid number of chromosomes. INCORRECT CHOICES: (2) In cells produced by mitosis, the chromosome number remains the same as the original number. (3) Fertilization restores the diploid number of chromosomes. (4) Polyploidy is a condition in which cells have twice the original number of chromosomes.

33. **(4)** The hypocotyl develops into the root and the lower part of the stem. INCORRECT CHOICES: (1) The epicotyl develops into the leaves and upper portions of the stem. (2) The cotyledon stores food. (3) The seed develops from the ovule.

34. **(1)** Asexual reproduction is any method of producing new organisms which does not involve fusion of nuclei; new organisms develop essentially by mitotic cell division. INCORRECT CHOICES: (2) Fertilization, or fusion of male and female nuclei, is sexual reproduction. (3) Formation of gametes, or functional sex cells, is associated with sexual reproduction. (4) Reduction of chromosome number as a result of meiosis occurs during the formation of gametes.

35. **(3)** Synapsis is the pairing of homologous chromosomes during the first meiotic division. INCORRECT CHOICES: (1) After synapsis, the homologous pairs become aligned on the spindle as pairs. (2) Disjunction refers to the separation of the members of an homologous pair of chromosomes as they move to opposite ends of the spindle. (4) Crossing-over involves the exchange of portions of one chromatid for corresponding portions of another chromatid during synapsis and disjunction.

36. **(3)** Budding is essentially mitotic division which is characterized by the unequal division of the cytoplasm. INCORRECT CHOICES: (1) Since budding is a form of asexual reproduction, fertilization, which represents sexual reproduction, could not occur. (2) Equal division is not a characteristic of budding. (4) Again, fertilization is a sexual term and it is used here in conjunction with asexual reproduction or mitosis.

37. **(1)** During the period prior to birth (gestation), the embryo develops in the uterus. INCORRECT CHOICES: (2) The egg develops in a follicle. (3) At the end of the period of gestation, the baby leaves the uterus through the vagina or birth canal. (4) The ovary produces the egg cells.

38. **(1)** The 2n or diploid number results from fertilization or the fusion of two sex cells. INCORRECT CHOICES: (2) and (4) The monoploid number or n number of chromosomes are formed in gamete formation or meiosis. (3) The diploid or 2n number of chromosomes exists throughout cleavage.

39. **(2)** The allantois functions in respiration and excretion. INCORRECT CHOICES: (1) The amnion encloses the amniotic fluid which provides a water environment and protection from shock. (3) The placenta is an adaptation for internal development in certain mammals. (4) The yolk sac contains blood vessels which make possible the transport of food to the developing chick.

40. **(3)** An ovule is the structure from which the seed develops. A seed is a ripened ovule which contains the embryo. An embryo is formed by the union of an ovum and sperm cell (fertilization). INCORRECT CHOICES: (1) The stem serves mainly for support and the conduction of materials up to the leaves and down to the roots. (2) A stamen is the male reproductive part of a flower. (4) The anther is the part of the stamen that bears the pollen grain.

41. **(2)** The function of the anther is to produce pollen grains by mitotic division. INCORRECT CHOICES: (1) The ovary is the female part of the reproductive structure. Pollen grains are male reproductive structures. (3) The stigma is the part of the pistil that receives the pollen grairn. (4) The root is a structure that anchors the plant, and absorbs nutrients and water from the soil.

42. **(2)** Gametes are monoploid in number. Therefore, when two gametes unite, the two monoploid numbers form one diploid number. INCORRECT CHOICES: (1) If zygotes unite to restore the diploid number, the chromosome number would be $4n$ in that each zygote is diploid. (3) Body cells are also diploid and do not take part in sexual reproduction. (4) For the reasons stated above these answers could not restore the diploid number.

43. **(4)** In both mitotic and meiotic cell division, replication of chromosomes occurs prior to the visual changes which mark these processes. INCORRECT CHOICES: (1) Neither mitosis nor meiosis produces diploid gametes from monoploid cells. (2) Meiotic cell division produces monoploid gametes from diploid cells. (3) Synapsis, or pairing, of homologous chromosomes occurs during the first meiotic division only.

44. **(2)** The monoploid number of chromosomes are found in the sex cells of organisms. INCORRECT CHOICES: (1), (3), and (4) The leaf bud, root hair, and seeds are somatic and have the diploid number of chromosomes in their cells.

45. **(3)** Stamens are the male reproductive organs which produce pollen grains. INCORRECT CHOICES: (1) and (2) Fruits develop from the pistil, the female reproductive organ. (4) Ovules are part of the female reproductive organ.

46. **(1)** The normal result of mitosis is the formation of two daughter nuclei which are identical to the original nucleus in number and types of chromosomes. INCORRECT CHOICES: (2) and (4) Each daughter cell of mitosis has the same number of chromosomes as the mother cell. (3) Meiosis produces daughter cells each of which has half as many chromosomes as the mother cell.

47. **(4)** To begin flower fertilization or development, there must be the process of pollination whereby the pollen attaches to the sticky stigma and forms the pollen tubes through which the sperm nuclei travel. The rest of the sequence is in order; fertilization of the ova by the sperm nuclei, seed formation within the ovary, and finally seed germination. INCORRECT CHOICES: (1) This choice refers to the development of fruits rather than flowers. (2) and (3) These choices refer to the sexual fertilization and development in animals.

48. **(1)** In looking at the diagram, we find that process A is at the end of the multicellular diploid adult and before the gametes. The logical assumption then is that process A is the formation of gametes, and in the human female this would occur in the ovary. INCORRECT CHOICES: (2) The uterus is the site for the implantation of the zygote. (3) The placenta is formed half from the embryo and half from the mother. It is designed to facilitate the absorption from the mother of nutrients and oxygen and the removal from the embryo of metabolic wastes. (4) The vagina functions as the birth canal and receptacle for the penis.

49. **(3)** The frog or amphibian lives some of its life in the water. It relies on the water to complete its life cycle. Process B, which signifies mitosis or growth of body cells in the frog, occurs in water and is the tadpole stage. INCORRECT CHOICES: (1) The frog's ovary merely produces the eggs which are then deposited in the water to await development. (2) The oviduct collects the eggs from the ovary and deposits them in the cloaca. (4) Since frogs are dependent on the water to complete their reproduction, this answer would not make much sense

50. **(4)** The cotyledon contains stored food in the form of starch. INCORRECT CHOICES: (1) Within pollen grains is found the sperm nucleus. (2) The leaves and upper part of the stem develops from the epicotyl. (3) The hypocotyl produces the roots and lower stem.

51. **(1)** Cell A is a cell in which cytoplasmic division is unequal, as in budding. INCORRECT CHOICES: (2) and (3) Cell B does not represent budding as the division of the cytoplasm is equal, as in fission. (4) Cell A does represent

a budding cell, as one newly formed cell has far less cytoplasm than the other.

52. **(1)** Hermaphrodites are organisms which contain functional male and female gonads, therefore, they can produce both eggs and sperm. IN-CORRECT CHOICES: (2) The production of eggs that develop without fertilization is parthenogenesis. (3) Asexual reproduction is any method of producing new organisms that does not involve fusion of nuclei. (4) Budding is a form of asexual reproduction.

53. **(1)** The ectoderm or outer skin layer does give rise to the epidermis and the nervous system. This is substantiated through studies in histology and embryology. INCORRECT CHOICES: (2) The blastula is one stage in the development of the zygote. (3) The mesoderm gives rise to structures like the skeleton, muscles, circulatory system, etc. (4) The gastrula is also a stage in the development of the embryo.

54. **(3)** The pancreas is least likely to affect the menstrual cycle in that this organ deals with digestion via its secretion of pancreatin fluid, and with regulation in its secretion of insulin. INCORRECT CHOICES: (1) The pituitary initiates the onset of the menstrual cycle by secreting FSH. (2) The ovary, in response to FSH initiates the growth of the follicle which in turn secretes estrogen. (4) The corpus luteum secretes progesterone which further develops the lining of the uterus.

55. **(4)** Cleavage is that series of mitotic cell divisions which leads to the formation of the animal embryo known as a blastula. INCORRECT CHOICES: (1) Parthenogenesis is the development of an unfertilized egg. (2) Meiosis is a form of nuclear division which results in the formation of monoploid gametes. (3) Reduction division is another term for meiosis.

56. **(2)** The question is asserting the birth of identical twins which indicates the fertilization of one egg and one male birth which we are to assume is not identical to the other. This would then mean the fertilization of the second egg. INCORRECT CHOICES: (1), (3), and (4) These choices would be incorrect following the logic from above.

57. **(2)** Grafting, a form of vegetative propagation, produces new cells as a result of mitosis; therefore, the new growth is genetically like that of the parent and will have its desirable characteristics. INCORRECT CHOICES: (1) Seeds develop as a result of fertilization and the plants which grow from them may display characteristics of both parents; furthermore, the development of a seedling into a tree is slow. (3) Cross-pollination will result in the formation of a seed with genes from both parents. (4) Highly

differentiated cells have usually lost their ability to divide and therefore cannot be cultured.

58. **(2)** Undifferentiated, meristematic tissue can produce new growth and is found in the cambium of the stem and the tip of the root. INCORRECT CHOICES: (1) The epidermis is the outer tissue of young root or stem. (3) Phloem tubes primarily transport sugar downward in the stem. (4) Xylem cells carry water and minerals upward.

59. **(4)** Meristems contain undifferentiated cells which undergo active cell reproduction; growth is restricted to these regions. INCORRECT CHOICES: (1) Stomates are openings in the leaf which permit the exchange of gases between the leaf and the atmosphere. (2) Amnions are embryonic membranes which help provide a favorable environment for development of certain animals. (3) Endoderms are among the three embryonic layers in multicellular animals.

60. **(4)** In humans the amnion protects the embryo from mechanical shock by its containment of amniotic fluid. INCORRECT CHOICES: (1) The chorion lines the inside of the shell in birds and attaches to the uterine wall in mammals. (2) The placenta is a means of transport for nutrients to the embryo and wastes from the embryo to the mother. (3) The allantois is a fetal membrane in birds which aids in the excretion of metabolic wastes and in respiration.

61. **(1)** In the diagram, number 1 is pointing to the epicotyl which does develop into the upper part of the stem. INCORRECT CHOICES: (2) Number 2 is pointing to the hypocotyl which will develop into the lower portion of the stem and roots. (3) Number 3 is pointing to the cotyledons which form the bulk of the embryo and are the seed leaves. (4) Number 4 is pointing to the seed coat or testa.

62. **(3)** The cotyledons do provide food for the developing embryo. INCORRECT CHOICES: (1) , (2) , and (4) Refer to number 61 above.

63. **(1)** The reduction of chromosome number from diploid to monoploid is characteristic of the process of meiosis. INCORRECT CHOICES: (2) Mitosis is the exact duplication of the type and number of chromosomes and the subsequent division into two daughter cells. (3) Polyploidy results in cells with one or more extra sets of chromosomes; $3n$, $4n$, etc. (4) Fertilization is the restoration of the $2n$ number.

64. **(1)** Monoploid means one half the normal amount of chromosomes. The process resulting in this reduction of chromosomes is meiosis. INCORRECT

CHOICES: (2) Mitosis results in the exact duplication of the type and number of chromosomes. (3) Fertilization is the union of the egg and sperm cell to form a diploid zygote. (4) Fission is simply the division, through mitosis, of the cell.

65. **(3)** Hermaphroditism is characterized by both sexes being present in the same organism. This choice is correct in that it states this fact. INCORRECT CHOICES: (1) Metamorphosis can be either complete or incomplete. It is a change in the anatomy of organisms as they develop. The change is brought about by hormones. (2) Hermaphroditism is not a prerequisite for terrestrial existence. (4) The earthworm carries on external fertilization and development.

66. **(2)** Number 6 is pointing to the yolk sac which provides for the nourishment of the embryo until birth. INCORRECT CHOICES: (1) Number 5 is pointing to the head of the embryo. (3) Number 3 is pointing to the amnion which protects the embryo from mechanical shock. (4) Number 4 is pointing to the amniotic chamber filled with amniotic fluid.

67. **(4)** The amniotic fluid would be found in the amniotic chamber which is indicated by this choice. INCORRECT CHOICES: (1), (2), and (3) Refer to number 66 above.

68. **(4)** The cell plate forms between the two newly formed nuclei as the plant cell divides in two. INCORRECT CHOICES: (1) This diagram shows two animal cells in which division of the cytoplasm was accomplished by a furrowing of the cytoplasm. (2) This diagram shows double-stranded chromosomes at the center of the spindle prior to the separation of single-stranded chromosomes. (3) This diagram shows the disjunction of homologous chromosomes during the first meiotic division.

69. **(3)** In this diagram, three pairs of homologous chromosomes can be seen, each pair consisting of two double-stranded chromosomes. INCORRECT CHOICES: (1) Individual chromosomes cannot be seen in the nuclei in these cells. (2) Here, six double-stranded chromosomes are attached to the spindle. (4) Individual chromosomes cannot be seen clearly in the nuclei of these cells.

70. **(2)** The picture shows an outgrowth forming on the parent cell. This is well recognized as budding, as seen in yeast cells. INCORRECT CHOICES: (1) Spore formation is characterized by bread mold and the picture would have to include sporangia, which it does not. (3) Fertilization is sexual and pictured here is asexual reproduction. (4) Seed formation is part of

the flower's sexual reproduction process. Again, this picture is showing asexual reproduction.

71. **(2)** Sperm cells are male sex cells. INCORRECT CHOICES: (1) Egg cells are female sex cells. (3) Testes are the male sex organs which produce male gametes. (4) Zygotes or fertilized eggs are formed by the union of male and female gametes.

72. **(3)** Soil is not needed for germination of seeds; they can germinate in other media. INCORRECT CHOICES: (1) , (2) , and (4) Sufficient oxygen, moisture, and a proper temperature are all conditions necessary for the germination of a seed.

73. **(4)** The key to the question is the term fission. Simple mitotic division always results in the same number of chromosomes and smaller cytoplasmic volume. INCORRECT CHOICES: (1) Having one half the number of chromosomes would imply meiosis and fission is not meiosis. (2) Having twice as many chromosomes would indicate polyploidy and sexual reproduction. Fission again is asexual. (3) This choice does not hold to the definition of mitosis.

74. **(3)** Homologous chromosomes are two chromosomes of the same pair; gametes contain only one chromosome of each pair. INCORRECT CHOICES: (1) Zygotes or fertilized eggs have two chromosomes of each pair. (2) In body cells, chromosomes are always found in pairs. (4) Even in the embryo, all body cells contain two chromosomes of each homologous pair.

75. **(1)** A cell plate is a structure which divides the cytoplasm of plants after division of the chromosomes. INCORRECT CHOICES: (2) Spindle fibers are produced by all cells. (3) A centromere connects chromatids in both plant and animal cells. (4) Chromosomes separate or undergo disjunction during mitosis in all cells.

76. **(2)** The embryo plant is formed in the ovule of the ovary. INCORRECT CHOICES: (1) A fruit is a ripened ovary adapted to disperse seeds. (3) A spore is a special cell, produced asexually, which can develop into a new organism. (4) A seed consists of the embryo and its seed coats.

77. **(3)** The oval body found in the middle of a pair of chromatids is the centromere. It holds the replicated chromosomes together. INCORRECT CHOICES: (1) Centrioles, found in animal cells, function in cell division. (2) Polar bodies are nonfunctional ova formed in oogenesis. (4) Spindle fibers connect (through the centromere) the chromatids with the polar regions.

78. **(3)** As the pollen tube migrates down the style, the sperm nuclei are formed. INCORRECT CHOICES: (1) The testes produce sperm cells in animals. (2) The ovules in the flowering plant produce eggs. (4) Polar bodies are made during the formation of the egg cell in animals.

79. **(3)** The question is referring to the sexual reproduction of flowering plants. The pollen grain is the male gamete and the ova within the ovary is the female gamete. INCORRECT CHOICES: (1) Pollination is the process of spreading the pollen either by wind, insects, or other means. (2) Parthenogenesis is the development of an egg without fertilization. (4) Cleavage is the rapid series of divisions that a zygote undergoes.

80. **(4)** Monoploid numbers are found in the gametes. The only answer that fits the label of a gamete is the pollen grain. INCORRECT CHOICES: (1) A root hair is a diploid root cell. It increases surface absorption, (2) A phloem cell is also diploid and its function is the translocation of food in plants. (3) An embryo, the result of fertilization, is diploid.

81. **(4)** Gametes or sex cells unite in sexual reproduction. INCORRECT CHOICES: (1) Fission or splitting in half is a form of asexual reproduction. (2) Budding is a type of asexual reproduction in which the cytoplasm divides unequally. (3) Vegetative propagation involves using the roots, stems, or leaves of plants for asexual reproduction.

82. **(4)** Grafting, a form of artifical propagation, is actually the attachment of different species (scion) to a parent tree (stock). The scion and stock cambium layers must be in contact for new cell growth to occur. INCORRECT CHOICES: (1) Bulb formation is a natural vegetative method of propagation. It involves one bulb, which through mitotic division gives rise to identical plants. (2) Fission is simple mitosis again resulting in identical offspring. (3) Spore formation is asexual reproduction which again gives rise to identical offspring.

83. **(4)** Cleavage is the early stage of development of the zygote or fertilized egg; it is characterized by rapid cell division without growth. INCORRECT CHOICES: (1) Postnatal development follows birth. (2) The membranes surrounding the embryo form after implantation in the uterus. (3) The placenta develops after the stages depicted in the diagram.

5

Transmission of Traits From Generation to Generation (Genetics)

A. Mendelian Heredity
   1. Dominance
   2. Segregation
   3. Independent assortment

B. The Gene-Chromosome Theory
   1. Dominance
   2. Segregation and recombination
   3. Independent assortment and recombination

C. Patterns of heredity
   1. Incomplete dominance
   2. Sex determination
   3. Sex linkage
   4. Multiple alleles

D. Mutations
   1. Types
      a. Chromosomal mutations
         (1) Crossing-over
         (2) Nondisjunction
         (3) Polyploidy
         (4) Other chromosomal mutations
      b. Gene mutations
   2. Mutagenic agents

E. Interaction of Heredity and Environment

F. Heredity and Man
   1. Human genetics
      a. Phenylketonuria
      b. Sickle-cell anemia
      c. Blood groups
   2. Plant and animal breeding

G. Modern Genetics
   1. DNA as the hereditary material
      a. Evidences
      b. DNA structure
         (1) Nucleotides
         (2) Watson–Crick model
   2. Gene action
      a. DNA replication
      b. Gene control of cellular activities
         (1) RNA
         (2) DNA and RNA codes
         (3) Protein synthesis
         (4) The "one gene-one enzyme" hypothesis
         (5) Individuality of organisms as related to their DNA and RNA codes
   3. Gene mutations
      a. Characteristics
      b. Mechanisms
   4. Cytoplasmic inheritance

The following questions on Genetics have appeared on previous Regents Examinations.

Directions: Each question is followed by four choices. Underline the correct choice.

1. During synapsis in meiosis, portions of one chromosome may be exchanged for corresponding portions of its homologous chromosome. This process is known as
   (1) nondisjunction
   (2) polyploidy
   (3) crossing-over
   (4) hybridization

2. The site for the production of most enzymes within a cell is the
   (1) centrioles
   (2) nuclei
   (3) ribosomes
   (4) vacuoles

3. Which is the sugar component of a DNA nucleotide?
   (1) adenine
   (2) deoxyribose
   (3) glucose
   (4) phosphate

4. In a cell, the transfer of genetic information from DNA to RNA occurs in the
   (1) cell membrane
   (2) endoplasmic reticulum
   (3) nucleus
   (4) nucleolus

5. Corn plants that are grown in the dark will be white and usually much taller than genetically identical corn plants grown in light, which will be green and shorter. The most probable explanation for this is that the
   (1) corn plants grown in the dark were all mutants for color and height
   (2) expression of a gene may be dependent on the environment
   (3) plants grown in the dark will always be genetically albino
   (4) phenotype of a plant is independent of its genotype

6. What are the possible blood types of a man whose child has type A and whose wife has type B blood?
   (1) A or B
   (2) B or AB
   (3) A or AB
   (4) B or O

7. If a trait which is not evident in the parents appears in their offspring, the parental genotypes are most likely
   (1) pure recessive
   (2) monoploid
   (3) homozygous
   (4) heterozygous

8. Phenylketonuria (PKU) is an inherited condition characterized by feeblemindedness. The symptoms of the disease result from the inability to synthesize a single type of
   (1) enzyme
   (2) nutrient
   (3) blood cell
   (4) brain cell

9. Which principle of heredity was developed by Gregor Mendel?
   (1) incomplete dominance
   (2) multiple alleles
   (3) sex linkage
   (4) independent assortment

10. Two plants hybrid for a single trait are crossed. If 100 offspring are produced, what percent could be expected to be homozygous for the dominant trait?
    (1) 0%   (3) 50%
    (2) 25%   (4) 75%

11. What is the total number of chromosomes in a typical body cell of a person with Down's syndrome?
    (1) 22   (3) 44
    (2) 23   (4) 47

12. In guinea pigs, a rough coat (R) is dominant over a smooth coat (r). Over a period of several years, two guinea pigs produced many offspring. Half the offspring had rough coats and half had smooth coats. The parents' genetic makeup is best represented by
    (1) $RR \times Rr$ (3) $Rr \times rr$
    (2) $Rr \times Rr$ (4) $RR \times rr$

13. Down's syndrome is an inherited human defect which results in mental retardation. The most common cause of this defect is known to be
    (1) gene linkage   (3) crossing-over
    (2) nondisjunction   (4) sex linkage

14. When two four-o'clock plants are crossed, 48 pink four-o'clocks and 52 white four-o'clocks are produced. The phenotypes of the parents are
    (1) pink and white   (3) pink and pink
    (2) pink and red   (4) red and white

15. The condition known as Down's syndrome may result from
    (1) nondisjunction of chromosome pair number 21
    (2) disjunction of chromosome pair number 18
    (3) crossing-over between homologous chromosomes
    (4) alteration of a single base pair of DNA

16. An organism possessing two identical genes for a trait is said to be
    (1) heterozygous for the trait
    (2) hybrid for the trait
    (3) homozygous for the trait
    (4) incompletely dominant for the trait

17. In humans, sex is normally determined at fertilization by
    (1) one pair of sex chromosomes
    (2) 2 pairs of sex chromosomes
    (3) 11 pairs of autosomes
    (4) 22 pairs of autosomes

18. A certain species of plant produces blue flowers when the soil pH is above 7.0. However, when the soil pH is below 7.0, the flowers are pink. Which statement best explains this color change? •
    (1) Mutagenic agents can alter genotypes.
    (2) The environment influences gene action.
    (3) Polyploidy produces 2n gametes.
    (4) Chromosomal mutations produce color effects.

19. Normal base pairing in DNA is limited to adenine-thymine and
    (1) guanine-cytosine
    (2) guanine-adenine
    (3) cytosine-adenine
    (4) cytosine-thymine

20. A DNA nucleotide is composed of three parts. These three parts may be
    (1) phosphate, adenine, and thymine
    (2) phosphate, deoxyribose, and thymine
    (3) phosphate, glucose, and cytosine
    (4) adenine, thymine, and cytosine

21. Based on the gene chromosome theory, the law of independent assortment assumes that certain genes are
    (1) formed by chromosomal mutations
    (2) located on the same chromosome
    (3) formed in the cytoplasm
    (4) located on separate chromosomes

22. What are the basic structural units of a DNA molecule?
    (1) glucose molecules
    (2) amino acids
    (3) lipids
    (4) nucleotides

23. According to the Hardy-Weinberg principle, which would most likely upset the stability of the gene pool of a population?
    (1) maintaining a large population
    (2) geographic isolation of part of the population
    (3) a lack of mutations
    (4) random mating

24. Animal breeders often cross members of the same litter in order to maintain desirable traits. This procedure is know as
    (1) hybridization
    (2) inbreeding
    (3) natural selection
    (4) vegetative propagation

25. A study of a trait in a large number of families showed that 50% of the boys and none of the girls exhibited this trait. The gene for this trait is most likely
    (1) autosomal dominant
    (2) incompletely dominant
    (3) homozygous
    (4) sex-linked

26. Which condition is a mutation caused by a change in the number of chromosomes?
    (1) hemophilia
    (2) albinism
    (3) color blindness
    (4) Down's syndrome

27. Only red tulips result from a cross between homozygous red and homozygous white tulips. This illustrates the principle of
    (1) independent assortment
    (2) dominance
    (3) segregation
    (4) incomplete dominance

Directions: Base your answers to questions 28 through 31 on your knowledge of biology and the diagrams below. The diagram on the left represents a portion of a double-stranded DNA molecule. The diagrams at the right represent specific combinations of nitrogenous bases found in compounds that are transporting specific amino acids.

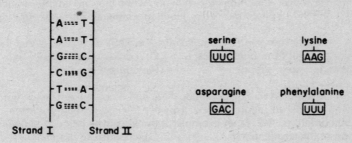

28. The amino acid whose genetic code is present in Strand I is
    (1) lysine
    (2) serine
    (3) asparagine
    (4) phenylalanine

29. The thymine (T) of Strand I is accidentally replaced by adenine (A). This occurrence is called
    (1) segregation
    (2) disjunction
    (3) cytoplasmic inheritance
    (4) gene mutation

30. The number of different amino acids coded by Strand I is
    (1) 1
    (2) 2
    (3) 8
    (4) 12

31. Which represents the sequence of nitrogenous bases in the molecule of messenger RNA synthesized by Strand I?
    (1) -T-T-C-G-U-C-
    (2) -A-A-C-G-T-C-
    (3) -U-U-C-G-A-C-
    (4) -A-A-G-C-U-G-

32. A man with type A blood whose mother had type O blood marries a woman with type O blood. What percentage of the offspring is expected to have type O blood?
    (1) 0%
    (2) 25%
    (3) 50%
    (4) 100%

33. Which is a major difference between messenger RNA molecules and transfer RNA molecules?
    (1) Messenger RNA molecules contain ribose, and transfer RNA molecules contain deoxyribose.
    (2) Messenger RNA molecules function in carrying coded information to the ribosomes, and transfer RNA molecules function in carrying amino acids to the ribosomes.
    (3) Messenger RNA molecules contain thymine, and transfer RNA molecules contain uracil.
    (4) Messenger RNA molecules function when they are double-stranded, and transfer RNA molecules function when they are single-stranded.

34. A similarity between DNA molecules and RNA molecules is that they
    (1) are built from nucleotides
    (2) are double-stranded
    (3) contain deoxyribose sugar
    (4) contain uracil

35. All the children of a hemophiliac male and a normal female are normal with respect to blood clotting. However, some of their grandsons are hemophiliacs. This is an example of the pattern of heredity known as
    (1) sex determination
    (2) sex-linkage
    (3) incomplete dominance
    (4) multiple alleles

36. Molecules which transport amino acids to ribosomes are known as
    (1) protein molecules
    (2) RNA molecules
    (3) mitochondria
    (4) chromosomes

Directions: Base your answers to questions 37 through 40 on the list of nucleic acid components below.

Some Nucleic Acid Components
(1) Ribose
(2) Deoxyribose
(3) Adenine
(4) Uracil
(5) Phosphate
(6) Thymine

37. Which components are found in both RNA and DNA molecules?
    (1) 1 and 2
    (2) 1 and 6
    (3) 3 and 5
    (4) 3 and 6

38. Which components may be present in RNA molecules, only?
    (1) 1 and 3
    (2) 1 and 4
    (3) 3 and 6
    (4) 4 and 6

39. In RNA molecules, the genetic code is made up of specific sequences of components. Examples of such components are
    (1) 1 and 2
    (2) 2 and 3
    (3) 3 and 4
    (4) 4 and 5

40. In DNA, which pair of components may be held together by relatively weak hydrogen bonds?
    (1) 1 and 5
    (2) 2 and 6
    (3) 3 and 5
    (4) 3 and 6

41. The specificity of genetic material is the result of the
    (1) type of sugar present in DNA
    (2) type of phosphate found in a cell
    (3) order of the nitrogen bases in DNA
    (4) order of the amino acids in a protein

42. The gene for normal pigmentation is dominant over its allele for albinism. Two parents with normal pigmentation have an albino child. What are the chances that their next child will be an albino?
    (1) 1 out of 4
    (2) 1 out of 2
    (3) 3 out of 4
    (4) 1 out of 1

43. Crosses between a certain species of red-flowered plants and white-flowered plants resulted in only pink-flowered offspring. What is the

expected percentage of flower color produced when these pink-flowered plants are crossed with each other?
(1) 100% pink
(2) 100% red
(3) 50% red, 50% white
(4) 25% white, 50% pink, 25% red

44. In which organelles are polypeptide chains synthesized?
(1) nuclei
(2) vacuoles
(3) ribosomes
(4) cilia

Directions: Base your answers to questions 45 through 47 on the diagram below which represents a segment of a DNA molecule and on your knowledge of biology.

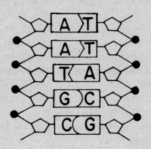

45. If the segment of DNA represented by the diagram was used as a template in the synthesis of messenger RNA, which sequence represents the order of bases found in the messenger RNA molecule?
(1) U-U-A-C-G
(2) T-T-A-G-C
(3) A-A-T-C-G
(4) T-T-U-G-C

46. This DNA molecule acts as a template for RNA construction in the process of
(1) gene replication
(2) protein synthesis
(3) osmosis
(4) synapsis

47. A change in the base sequence in this DNA molecule is known as
(1) homeostatic control
(2) gene segregation
(3) disjunction
(4) a gene mutation

48. The "one-gene, one-enzyme" hypothesis deals most directly with the relationship of genes to the synthesis of

(1) polypeptides

(3) lipids

(2) polysaccharides

(4) carbohydrates

49. A high concentration of an enzyme that breaks down RNA molecules is introduced into a cell. Which cellular activity would probably be affected first?
    (1) metabolism of fats
    (2) synthesis of proteins
    (3) hydrolysis of ATP
    (4) oxidation of glucose

50. Watson and Crick described the DNA molecule as a
    (1) straight chain
    (2) single strand
    (3) double helix
    (4) branching chain

51. A family has 3 boys and 1 girl. What is the chance that the next child will be a girl?
    (1) 25%
    (2) 50%
    (3) 75%
    (4) 100%

52. Evidence has begun to accumulate which suggests the presence of hereditary systems outside the cell nucleus. It appears that these new hereditary systems can be found in both
    (1) cell walls and cell membranes
    (2) nucleoli and vacuoles
    (3) ribosomes and centrosomes
    (4) mitochondria and chloroplasts

53. A gardener found that when white petunias and purple petunias were crossed, only blue petunias were produced. From which of the following crosses would the gardener most probably obtain the greatest percentage of white petunias?
    (1) white and blue petunias
    (2) purple and purple petunias
    (3) blue and blue petunias
    (4) purple and blue petunias

Directions (54–58): For each statement in questions 54 through 58, select the biological term, *chosen from the list below*, that is best described by that statement.

Biological Terms
(1) Sex-linked inheritance
(2) Inbreeding
(3) Mutation
(4) Vegetative propagation
(5) Incomplete dominance
(6) Dominance

54. Hybrid plants such as plumcots are maintained by grafting.

55. A cross between a black Andalusian fowl and a white Andalusian fowl results in a blue offspring.

56. When a hybrid black guinea pig is crossed with a pure black guinea pig, all of the offspring are black.

57. In humans, color blindness is associated with the X-chromosome.

58. Ionizing radiation can cause changes in DNA molecules.

59. An individual contains the sex chromosomes XYY. This combination of chromosomes was most likely the result of
    (1) DNA replication   (3) nondisjunction
    (2) cleavage   (4) crossing-over

60. When plants that produce round squash are crossed with plants that produce long squash, the F_1 generation produces only oval squash. Which pattern of heredity does this cross illustrate?
    (1) incomplete dominance   (3) dominance-recessiveness
    (2) sex-linkage   (4) independent assortment

Directions: Base your answers to questions 61 and 62 on the information below and your knowledge of biology.

A sample of amniotic fluid is collected for examination during the early part of embryonic development. The cells in the fluid are cultured and then the chromosomes are separated from the cells and observed under a microscope.

61. Which condition could be detected by this procedure?
    (1) color blindness   (3) hemophilia
    (2) Down's syndrome   (4) phenylketonuria

62. From which region of the reproductive system represented on the next page was the amniotic fluid obtained?

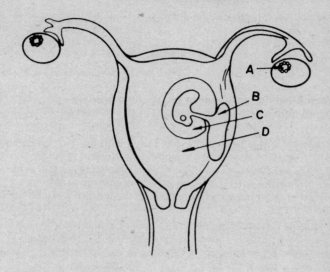

(1) A   (3) C
(2) B   (4) D

63. Parents with blood genotypes I^aI^b and I^bi would *not* produce a child with blood type
    (1) A   (3) AB
    (2) B   (4) O

Directions: Base your answers to questions 64 through 66 on the diagram below which represents some molecules involved in protein synthesis.

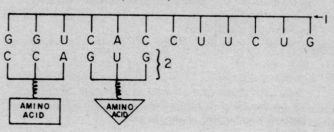

64. Strand 1 represents a molecule of
    (1) DNA
    (2) protein
    (3) transfer RNA
    (4) messenger RNA

65. In which part of the cell does the process represented by this diagram probably take place?
    (1) nucleus
    (2) nucleolus
    (3) ribosome
    (4) centrosome

66. Molecule 2 may be described as
    (1) an enzyme specific for a particular chemical reaction
    (2) a type of nucleic acid specific for a particular amino acid
    (3) a template used in the replication of nucleic acids
    (4) an amino acid which functions in the form of a double helix

67. A student tossed 2 pennies at the same time and recorded the following results: both tails, 23; one head and one tail, 53; both heads, 21. Which genotypes represent a cross resulting in approximately the same ratio?
    (1) AA × aa
    (2) Aa × aa
    (3) Aa × AA
    (4) Aa × Aa

68. The gene frequency of which trait could be studied by means of the Hardy-Weinberg principle?
    (1) chloroplast shape in mitotically dividing algae
    (2) feather color in sexually reproducing fowl
    (3) leg number in bees reproducing parthenogenetically
    (4) cilia length in fissioning Paramecia

69. In sexual reproduction the offspring may exhibit some traits different from either parent. Of the following, the most probable explanation is that
    (1) the parents were homozygous for most characteristics
    (2) dominant characteristics are the only ones that appear in the next generation
    (3) some pairs of genes for recessive characteristics were combined in producing the offspring
    (4) acquired characteristics of the hybrid parents were transmitted

Directions: Base your answers to questions 70 through 73 on the diagram on the next page which represents parts of two nucleic acid molecules.

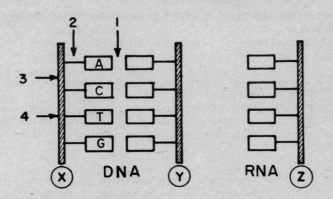

70. What is the normal nitrogenous base sequence in the segment of strand Y shown in the diagram?
    (1) T-G-A-C
    (2) T-A-A-C
    (3) U-G-A-C
    (4) T-G-U-C

71. Molecules represented by strand Z move from the nucleus of a cell to cytoplasmic organelles known as
    (1) mitochondria
    (2) vacuoles
    (3) ribosomes
    (4) centrioles

72. If strand X serves as a template for the synthesis of strand Z, the base sequence of that fragment of strand Z shown is
    (1) T-G-A-C
    (2) U-G-A-C
    (3) A-C-U-G
    (4) A-C-T-G

73. At which location does the DNA molecule "unzip" during replication?
    (1) 1
    (2) 2
    (3) 3
    (4) 4

74. Down's syndrome is a condition which occurs as a result of
    (1) crossing-over
    (2) polyploidy
    (3) gene mutation
    (4) nondisjunction

75. In order for a substance to act as a carrier of hereditary information, it must be
    (1) easily destroyed by enzyme action
    (2) exactly the same in all organisms
    (3) present only in the nuclei of cells
    (4) copied during the process of mitosis

76. The coded information in a DNA molecule directly determines the formation of
    (1) polypeptides
    (2) polysaccharides
    (3) lipids
    (4) monosaccharides

77. Recent investigations suggest that chloroplasts and mitochondria
    (1) do not contain genes
    (2) are completely controlled by genes located within their nuclei
    (3) contain separate genes which are regulated by genes in the centrosomes
    (4) contain separate genes which are regulated by genes in the nucleus

Directions: Base your answer to question 78 on the information below and on your knowledge of biology.

In a classroom survey, 50 students out of 200 were not able to taste PTC. The ability to taste PTC is a dominant genetic trait. [Use the Hardy-Weinberg equations to answer the following question.]

78. The frequency of the homozygous recessive genotype is
    (1) 40%
    (2) 33%
    (3) 25%
    (4) 4%

Directions: Base your answers to questions 79 through 81 on the graph below which represents the frequency of genotypes in a population.

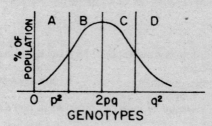

79. Which portions of the graph represent only heterozygous individuals?
    (1) *A* and *B*
    (2) *B* and *C*
    (3) *C* and *D*
    (4) *A* and *D*

80. Which portions of the graph represent only homozygous genotypes?
    (1) A and D
    (2) A and C
    (3) B and D
    (4) B and C

81. A natural disaster isolated group D from the rest of the population. Which would be the most immediate occurrence?
    (1) Group D would form a new species.
    (2) The mutation rate would be higher in D than in the remaining population.
    (3) The stability of the population would be altered.
    (4) The organisms of group A would migrate.

82. Which concept is most directly supported by the ability of chloroplasts to self-duplicate?
    (1) chromosomal nondisjunction
    (2) cytoplasmic inheritance
    (3) bacterial transformation
    (4) gene mutation

Directions: Base your answers to questions 83 through 85 on the information below which represents a change in a portion of the base sequence in a DNA molecule.

A ⊤ C G A T ─X-ray→ A Ⓐ C G A T

83. One important result of this type of change is that it
    (1) involves many chromosomes at once
    (2) may be passed on to offspring
    (3) prevents a change in an organism's phenotype
    (4) never affects an organism's genotype

84. This change can best be interpreted as
    (1) a gene mutation
    (2) nucleic acid replication
    (3) protein synthesis
    (4) gene replication

85. In humans, a change similar to the one shown has been responsible for a disorder known as
    (1) Down's syndrome
    (2) polyploidy
    (3) sickle-cell anemia
    (4) phagocytosis

86. Which cellular process involves DNA replication?
    (1) mitosis
    (2) cyclosis
    (3) pinocytosis
    (4) protein synthesis

87. Examples of self-duplicating cellular structures are the
    (1) mitochondria and chloroplasts   (3) cell walls and chloroplasts
    (2) mitochondria and cell walls   (4) vacuoles and chloroplasts

88. Studies have shown that the differences in identical twins are normally due to
    (1) having different genotypes
    (2) developing from two fertilized egg cells
    (3) being born at different times
    (4) being raised under different conditions

89. The ability of chloroplasts to replicate is an example of
    (1) enzyme replication   (3) autosomal dominance
    (2) cytoplasmic inheritance   (4) segregated genetics

90. A boy has brown hair and blue eyes, and his brother has brown hair and brown eyes. The fact that they have different combinations of traits is best explained by the concept known as
    (1) multiple alleles   (3) sex linkage
    (2) incomplete dominance   (4) independent assortment

91. A double-stranded DNA molecule replicates as it unwinds and "unzips" along weak
    (1) hydrogen bonds   (3) phosphate groups
    (2) carbon bonds   (4) ribose groups

92. What is the function of DNA molecules in the synthesis of proteins?
    (1) They catalyze the formation of peptide bonds.
    (2) They determine the sequence of amino acids in a protein.
    (3) They transfer amino acids from the cytoplasm to the nucleus.
    (4) They supply energy for protein synthesis.

Directions: Base your answers to questions 93 through 95 on the diagram on the next page which represents some functions of nucleic acids in a cell.

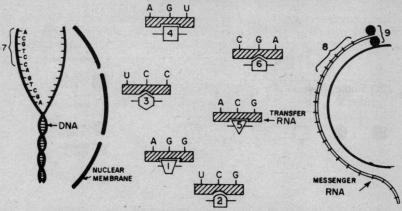

93. If structure number 7 of the DNA molecule in the diagram were to code structure number 8 of the messenger RNA, what would the first six letters in the code of the messenger RNA be?

(1)	(2)	(3)	(4)
T	U	A	U
G	G	C	C
C	C	G	G
A	A	T	U
G	G	C	G
G	G	C	G

94. The letters on the DNA and transfer RNA molecules represent
    (1) phosphates
    (2) nucleic acids
    (3) 5-carbon sugars
    (4) nitrogen bases

95. During the process represented in the diagram, messenger RNA becomes associated with structure number 9. This structure most likely represents a
    (1) ribosome
    (2) mitochondrion
    (3) nucleolus
    (4) centrosome

Directions: Base your answers to questions 96 through 100 on the pedigree chart on the next page which shows a family history of color blindness. In this sex-linked pattern of inheritance, the gene for normal vision (N) is dominant over the gene for color blindness (n).

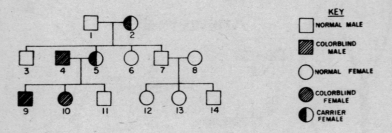

KEY

☐ NORMAL MALE

▨ COLORBLIND MALE

◯ NORMAL FEMALE

⬤ COLORBLIND FEMALE

◑ CARRIER FEMALE

96. What is the genetic makeup of individual 5?
    (1) $X^N X^N$ (3) $X^n X^n$
    (2) $X^N X^n$ (4) $X^n Y$

97. Which two individuals in the pedigree chart could be identical twins?
    (1) 5 and 6   (3) 9 and 11
    (2) 7 and 8   (4) 12 and 13

98. If female 6 married a colorblind male and had four children, what percent of their offspring will have normal vision?
    (1) 0%   (3) 50%
    (2) 25%   (4) 100%

99. If male 11 married a colorblind female, which will be true of the children that could be produced from this cross?
    (1) Only the sons will be colorblind.
    (2) Only the daughters will be colorblind.
    (3) All of the offspring will have normal vision.
    (4) All of the offspring will be colorblind.

100. Which is characteristic of both individuals 4 and 5?
    (1) Both individuals are colorblind.
    (2) Both individuals have normal vision.
    (3) Both individuals have one gene for color blindness.
    (4) Both individuals have two genes for color blindness.

Answers

1. 3	26. 4	51. 2	76. 1
2. 3	27. 2	52. 4	77. 4
3. 2	28. 1	53. 1	78. 3
4. 3	29. 4	54. 4	79. 2
5. 2	30. 2	55. 5	80. 1
6. 3	31. 3	56. 6	81. 3
7. 4	32. 3	57. 1	82. 2
8. 1	33. 2	58. 3	83. 2
9. 4	34. 1	59. 3	84. 1
10. 2	35. 2	60. 1	85. 3
11. 4	36. 2	61. 2	86. 1
12. 3	37. 3	62. 3	87. 1
13. 2	38. 2	63. 4	88. 4
14. 1	39. 3	64. 4	89. 2
15. 1	40. 4	65. 3	90. 4
16. 3	41. 3	66. 2	91. 1
17. 1	42. 1	67. 4	92. 2
18. 2	43. 4	68. 2	93. 2
19. 1	44. 3	69. 3	94. 4
20. 2	45. 1	70. 1	95. 1
21. 4	46. 2	71. 3	96. 2
22. 4	47. 4	72. 2	97. 4
23. 2	48. 1	73. 1	98. 4
24. 2	49. 2	74. 4	99. 1
25. 4	50. 3	75. 4	100. 3

Explanatory Answers

1. **(3)** As a result of crossing-over during meiosis, corresponding portions of homologous chromatids are exchanged, and linked genes are separated. INCORRECT CHOICES: (1) In nondisjunction, entire chromosomes, not portions of chromosomes, fail to separate. (2) Polyploidy results in

cells with one or more extra sets of chromosomes beyond the diploid number. (4) Hybridization is the mating of individuals in an attempt to bring together desirable combinations of genes.

2. **(3)** Since enzymes are protein in nature and ribosomes produce proteins, it would seem logical that enzymes are produced in the ribosomes. INCORRECT CHOICES: (1) The centrioles function in animal cell mitosis. (2) The nuclei control the cell's activities. (4) The vacuoles in cells are for storage, digestion, and excretion.

3. **(2)** DNA stands for deoxyribonucleic acid. The sugar component is a deoxygenated form of ribose. INCORRECT CHOICES: (1) Adenine is a nitrogenous base which composes DNA. (3) Glucose is a simple carbohydrate. (4) Phosphates are also components of DNA and compose the "ladders" of the double helix.

4. **(3)** DNA is part of the chromosomes in the nucleus and the synthesis of RNA takes place there. INCORRECT CHOICES: (1) Chromosomal DNA does not leave the nucleus and there is no DNA in the cell membrane. (2) Ribosomes along the endoplasmic reticulum are the sites where genetic information is translated from RNA into protein but there is no DNA there. (4) Messenger RNA which picks up genetic information from DNA is not synthesized in the nucleolus.

5. **(2)** The only variable factor described is the amount of light in the area in which the plants are grown. INCORRECT CHOICES: (1) The possibility of all being mutants is highly unlikely. (3) Albinism is inherited and not caused by environmental conditions. (4) The phenotype, or how a plant appears, is partially determined by the inherited traits or genotype.

6. **(3)** The gene for type A blood must come from the father; therefore, he must be type A or AB. INCORRECT CHOICES: (1) , (2) , and (4) A person with B or O type blood does not have a gene for A type blood.

7. **(4)** If both parents are heterozygous or hybrid, they have hidden, recessive genes which can combine and appear in their offspring. INCORRECT CHOICES: (1) Pure recessive parents will only produce offspring with recessive traits, (2) Monoploid cells have only one chromosome of each pair. (3) Homozygous parents will produce offspring with the same traits as their parents.

8. **(1)** An important enzyme which controls the breakdown of phenylalanine cannot be synthesized due to the lack of the required gene. INCORRECT CHOICES: (2) PKU is not caused by a single type of nutrient in

the diet. (3) Blood cells are synthesized in the bone marrow and are not affected by PKU. (4) Brain cells are damaged by the accumulation of phenylalanine, which cannot be broken down due to the absence of the required enzyme.

9. **(4)** As a result of crosses where two traits were studied at a time, Mendel developed the principle of independent assortment. INCORRECT CHOICES: (1) Incomplete dominance describes a situation in which heterozygous individuals do not resemble individuals with either of the homozygous combinations; Mendel studied situations involving complete dominance. (2) Multiple alleles refers to hereditary patterns where more than two alleles are involved in the inheritance of particular traits. (3) Sex-linkage describes situations in which the genes for particular traits are located on the X chromosome.

10. **(2)** Twenty-five percent of the offspring of a cross between hybrids can be expected to be homozygous dominant.

	A	a
A	AA	Aa
a	Aa	aa

INCORRECT CHOICES: (1) Twenty-five percent may be homozygous dominant, AA. (3) Fifty percent may be expected to be heterozygous, Aa. (4) Seventy-five percent may be expected to display the dominant phenotype.

11. **(4)** Down's syndrome results from the inheritance of an additional chromosome, number 21, resulting in a total of 47 chromosomes. INCORRECT CHOICES: (1) and (3) These do not represent the numbers of chromosomes in a human sex or somatic (body) cell. (2) There are 23 or the monoploid number of chromosomes in the human sperm or egg cell.

12. **(3)** A cross between a heterozygous rough (Rr) and a homozygous smooth (rr) can be expected to produce offspring half of whom are rough and half smooth.

	R	r
r	Rr	rr
r	Rr	rr

INCORRECT CHOICES: (1) All the offspring of the cross RR × Rr will have rough coats. (2) Seventy-five percent of the offspring of the cross Rr × Rr can be expected to have rough coats. (4) All the offspring of the cross RR × rr will have rough coats.

13. **(2)** Nondisjunction involves the failure of a pair of chromosomes to separate during disjunction; Down's syndrome results from nondisjunction of chromosome pair 21. INCORRECT CHOICES: (1) Nonallelic genes located on the same chromosome are said to be linked. (3) Crossing-over involves the exchange of portions of one chromatid for corresponding chromatids on homologous chromosomes during synapsis in the first meiotic division. (4) Genes located on the X chromosome are said to be sex-linked.

14. **(1)** A 1:1 ratio (48 pink: 52 white) may be expected when one parent is heterozygous and the other homozygous; four o'clocks exhibit incomplete dominance for flower color.

	R	W
W	RW	WW
W	RW	WW

    RW = pink phenotype, WW = white.

    INCORRECT CHOICES: (2) A cross between a pink (RW) four o'clock and a red (RR) would produce an expected ratio of one pink to one red among the offspring. (3) Two pink four o'clocks would produce an expected ratio of one red to two pinks to one white. (4) A cross between a red (RR) and a white (WW) would produce all pink offspring.

15. **(1)** Down's syndrome is a result of nondisjunction, which is the failure of chromosomes to separate after synapsis. This particular nondisjunction does occur on chromosome number 21. INCORRECT CHOICES: (2) Disjunction would mean the separation of the chromosome after synapsis. This, then, would be the opposite of nondisjunction. (3) Crossing-over involves the exchange of genetic information between the chromosomes forming a tetrad. (4) An alteration of a single base pair would affect the production of the correct protein; it might affect a single trait, but it would not affect the ability of chromosomes to separate after synapsis.

16. **(3)** When the two genes of an allelic pair are identical, the individual with these genes is said to be homozygous. INCORRECT CHOICES: (1) When the two genes of an allelic pair are unlike, the individual with these genes is said to be heterozygous. (2) When an individual possesses contrasting alleles, the individual may also be said to be hybrid (heterozygous). (4) Incomplete dominance involves a situation in which heterozygous individuals (which possess two unlike genes) do not resemble individuals with either of the homozygous combinations.

17. **(1)** The sex of a human is normally determined at fertilization by the random combination of the X and Y chromosomes; XX is the condition in females, XY in males. INCORRECT CHOICES: (2) Each pair of chromosomes including the sex chromosomes is divided by meiosis into the gametes so that each gamete carries one chromosome of a pair; fertilization recombines the chromosomes to produce a pair of each kind. (3) and (4) Autosomes are chromosomes not involved with the determination of sex; human cells carry 22 pairs of autosomes.

18. **(2)** The environment, here soil pH, interacts with genes in the expression of inherited traits. INCORRECT CHOICES: (1) Mutagenic agents produce heritable changes in the genes, but in this plant species the genes affecting flower color can produce blue flowers only at a pH above 7.0. (3) Polyploid individuals often exhibit exaggerated characteristics regardless of environmental conditions. (4) Chromosomal mutations usually involve many genes and some effects would be visible in the phenotype of the organism regardless of environment.

19. **(1)** Guanine–cytosine is a normal base pair found in DNA. INCORRECT CHOICES: (2) Guanine pairs only with cytosine and adenine pairs only with thymine. (3) Cytosine pairs only with guanine and adenine with thymine. (4) Cytosine pairs only with guanine and thymine with adenine.

20. **(2)** A DNA nucleotide is composed of a phosphate, the sugar deoxyribose, and a nitrogen base. INCORRECT CHOICES (1) and (4) There is no sugar molecule present. (3) Glucose is not the sugar present in DNA.

21. **(4)** Genes located on separate chromosomes are inherited independently of each other. INCORRECT CHOICES: (1) A chromosomal mutation is a change in chromosome number or structure. (2) Genes located on the same chromosome are linked, not inherited independently. (3) Genes are located in the nucleus.

22. **(4)** DNA is a nucleic acid made up of basic units called nucleotides. INCORRECT CHOICES: (1) Glucose molecules are the building blocks of carbohydrates. (2) Amino acids are the structural units of proteins. (3) Lipids are either fats or oils.

23. **(2)** Geographic isolation would be a contrast to the idea of random migrations which the principle states. INCORRECT CHOICES: (1) , (3) , and (4) All of these choices refer to statements included within the principle.

24. **(2)** Inbreeding is the crossing of organisms with the same desired inheritable traits to attempt to keep those traits in the offspring. INCORRECT

CHOICES: (1) Hybridization is the crossing of individuals with different genes for the same trait resulting in a variety within a species. (3) Natural selection is a process where those specimens best adapted to an environment survive. (4) Vegetative propagation is the formation of additional offspring from body or somatic cells of an organism, not allowing for inherited variations.

25. **(4)** A sex-linked gene is located on the X chromosome and is therefore linked to the genes that determine sex. INCORRECT CHOICES: (1) An autosomal dominant gene is located on one of the autosomes (nonsex chromosomes) and when present in a heterozygote is dominant over its recessive allele. (2) Incompletely dominant genes do not mask or dominate their alleles when present in heterozygotes. (3) Homozygous individuals carry two identical genes of an allelic pair.

26. **(4)** Down's syndrome results from nondisjunction of chromosome number 21 and a consequent 47 chromosomes in the afflicted individual. INCORRECT CHOICES: (1) and (3) Hemophilia and color blindness are caused by two separate but sex-linked, recessive genes. (2) Albinism is caused by an autosomal recessive gene which is expressed only in the homozygous condition.

27. **(2)** Dominance is illustrated by a cross in which the offspring of two parents having alternate forms of a contrasting character, here red and white, resemble only one parent, here the dominant red one. INCORRECT CHOICES: (1) Independent assortment refers to the independent inheritance of two separate traits. (3) Segregation refers to the reappearance of the hidden recessive character in the F_2 generation. (4) Incomplete dominance involves an apparent "blending" of the effects of contrasting alleles; if incomplete dominance were involved here, the offspring of red and white parents might be pink.

28. **(1)** On Strand I, lysine is encoded by the three bases AAG. INCORRECT CHOICES: (2) Serine (UUC) is not found on Strand I and uracil is not found in a DNA code. (3) Asparagine (GAC) is not found on Strand I. (4) Phenylalanine (UUU) is not found on either strand and uracil is not found in a DNA code.

29. **(4)** Any change in a base sequence in a DNA molecule is a gene mutation. INCORRECT CHOICES: (1) Segregation occurs when genes separate during meiosis. (2) Disjunction is the separation of entire chromosomes during meiosis. (3) Cytoplasmic inheritance occurs when certain organelles in the cytoplasm reproduce themselves.

30. **(2)** Three bases code one amino acid. There are six bases resulting in two amino acids in Strand I. INCORRECT CHOICES: (1) If there are two groups of three bases in Strand I, then more than one amino acid is coded. (3) and (4) Only two groups of three bases are found in Strand I, therefore, there are no more than two amino acids coded.

31. **(3)** In forming a molecule of RNA, T pairs with U, G with C, and C with G. INCORRECT CHOICES: (1) and (2) RNA does not contain the base T. (4) RNA is formed according to the base pairing rule whereby A pairs with U, G with C, and T with A, not U.

32. **(3)** Fifty percent of the offspring of this marriage is expected to have type O blood.

	I^a	i
i	$I^a i$	ii
i	$I^a i$	ii

INCORRECT CHOICES: (1) Zero percent of the offspring of this marriage will have type B or type AB blood. (2) Twenty-five percent is not an expected percentage among the offspring of this marriage. (4) One hundred percent of the offspring with type A blood will be heterozygous.

33. **(2)** Messenger RNA carries coded information from DNA to the ribosomes while transfer RNA carries amino acids to the ribosomes. INCORRECT CHOICES: (1) All types of RNA contain ribose; only DNA contains deoxyribose. (3) No type of RNA contains thymine while uracil may be found in all types of RNA. (4) Both messenger RNA and transfer RNA are single-stranded molecules.

34. **(1)** Both DNA and RNA are nucleic acids composed of nucleotides. INCORRECT CHOICES: (2) RNA is not double stranded. It is single stranded. (3) RNA does not contain deoxyribose sugar. It does contain ribose sugar. (4) DNA does not contain uracil.

35. **(2)** When an inherited trait is exhibited by only males or only females, it is said to be sex-linked. INCORRECT CHOICES: (1) The sex of an offspring is determined by the gene for sex that it inherits from the father; X for female and Y for male. (3) When the appearance of an organism reflects a melding or combining of two traits, those traits are said to have incomplete dominance over one another. (4) Hemophilia is not caused by having more than two genes for a trait or multiple alleles.

36. **(2)** Transfer RNA molecules bring specific amino acids to the ribosomes

to form polypeptides which form proteins. INCORRECT CHOICES: (1) Protein molecules are formed at the site of the ribosomes. (3) Mitochondria are organelles which regulate the release of energy from glucose. (4) Chromosomes are composed of DNA which codes the inherited traits.

37. **(3)** Adenine is a base found in both RNA and DNA; phosphate molecules are present in all nucleic acids. INCORRECT CHOICES: (1) Ribose is a sugar which is not found in DNA; deoxyribose is a sugar not found in RNA. (2) Ribose is not found in DNA. (4) Thymine is not found in RNA.

38. **(2)** Ribose is a sugar and uracil is a nitrogen base present only in RNA. INCORRECT CHOICES: (1) Adenine is a nitrogen base present in both RNA and DNA. (3) Adenine is found in both RNA and DNA; thymine is not found in RNA. (4) Uracil is substituted for thymine in RNA.

39. **(3)** The bases, adenine and uracil, may be used in the RNA genetic code. INCORRECT CHOICES: (1) The sugars, ribose and deoxyribose, are not components used in the genetic code. (2) Deoxyribose is not found in RNA; adenine may be used in the genetic code. (4) In RNA, uracil can be used to code traits, but the phosphate molecule is never used in coding.

40. **(4)** Nitrogen bases, such as adenine and thymine, are held together by hydrogen bonds. INCORRECT CHOICES: (1) Ribose is not present in DNA. (2) and (3) Only bases are held together by hydrogen bonds; deoxyribose and phosphate are not bases.

41. **(3)** The sequence of nitrogen bases in DNA acts as a code which determines the inheritance of specific traits. INCORRECT CHOICES: (1) Only one sugar, deoxyribose, is present in DNA. (2) There is only one form of phosphate found in nucleic acids. (4) The order of amino acids determines the specificity of proteins.

42. **(1)** Two normal parents who have an albino child must both be heterozygous, therefore, the chances that their next child will be an albino are 1 out of 4.

	N	n
N	NN	Nn
n	Nn	nn

nn is the genotype of an albino.

INCORRECT CHOICES: (2) The chances that their next child will be heterozygous are 1 out of 2. (3) The chances that their next child will have normal pigmentation are 3 out of 4. (4) 1 out of 1 equals 100% or absolute certainty; there can be no absolute certainty about which genes the children of these parents will receive.

43. **(4)** Pink flower color is the result of incomplete dominance between genes *R* and *W*; when pink (*RW*) is crossed with pink (*RW*), the expected percentage of flower color is 25% white (*WW*), 50% pink (*RW*), 25% red (*RR*).

	R	W
R	RR	RW
W	RW	WW

INCORRECT CHOICES: (1) Because pink is the result of the heterozygous condition, a cross between two pinks may produce red and white as well as pink. (2) Since each pink plant carries a gene for white (*W*) as well as one for red (*R*), it is unlikely that all the offspring would inherit two genes for red. (3) The offspring of a cross between two pinks will be expected to be 50% pink.

44. **(3)** Ribosomes are organelles at which amino acids are bonded in sequence to form polypeptide chains. INCORRECT CHOICES: (1) Nuclei are organelles which contain chromosomes. (2) Vacuoles are organelles which store materials. (4) Cilia are hairlike projections of protoplasm.

45. **(1)** Assuming the left side of the diagrammed molecule to act as the template, then the messenger RNA molecules will look like the right side but with *U* (uracil) replacing *T* (thymine). INCORRECT CHOICES: (2), (3), and (4) *T* (thymine) is never part of an RNA molecule.

46. **(2)** DNA transfers its genetic code to messenger RNA which then moves to the ribosomes where amino acids are bonded to form polypeptides in a sequence determined by the code carried by the messenger RNA; proteins are composed of one or more polypeptide chains. INCORRECT CHOICES: (1) Gene replication involves the synthesis of DNA using existing DNA as a template. (3) Osmosis is the diffusion of water through a semipermeable membrane. (4) Synapsis refers to the pairing of homologous chromosomes during meiosis.

47. **(4)** A gene mutation may be interpreted biochemically as any change that affects the base sequence in the DNA molecule. INCORRECT CHOICES: (1) Homeostatic control refers to the regulation of body processes so as to maintain the stability of the internal environment. (2) Gene segregation

refers to the separation of allelic genes during meiosis. (3) Disjunction describes the movement apart of chromosomes during mitosis and meiosis.

48. **(1)** A gene is the portion of a chromosome which codes a polypeptide; enzymes are made up of polypeptides. INCORRECT CHOICES: (2), (3), and (4) Polysaccharides, lipids, and carbohydrates are not protein in nature and cannot be building blocks of enzymes.

49. **(2)** DNA codes the synthesis of proteins. RNA components are needed to transfer the code to the ribosomes, where proteins are made. INCORRECT CHOICES: (1) and (4) RNA is not specifically required for the breakdown of fats and the oxidation of glucose. (3) The hydrolysis of ATP is controlled by ATP-ase.

50. **(3)** The DNA molecule, according to the Watson–Crick model, is double-stranded and twisted into the form of a helix. INCORRECT CHOICES: (1) DNA is not a straight chain but rather is coiled into a helical shape. (2) Each DNA molecule consists of two chains of nucleotides, that is, DNA is double-stranded. (4) The DNA molecule has no branches.

51. **(2)** Sex determination is based on the Punnett square showing the probability of two boys to two girls or 50%. INCORRECT CHOICES: (1) Twenty-five percent does not prove out using a Punnett square. (3) and (4) Again these percentages do not show on a Punnett square for sex determination.

52. **(4)** Mitochondria and chloroplasts have been found to contain their own unique DNA. INCORRECT CHOICES: (1) Neither cell walls nor cell membranes have been found to contain hereditary systems of their own. (2) Nucleoli are found within the nucleus; vacuoles do not possess hereditary systems of their own. (3) The inheritance of ribosomes and centrosomes is under the control of the cell nucleus.

53. **(1)** This cross is an example of incomplete dominance. Crossing white (*WW*) petunias with blue (*WP*) produces 50% white (*WW*). INCORRECT CHOICES: (2) Purple petunias (*PP*) crossed with purple (*PP*) yields 100% purple (*PP*). (3) Blue (*WP*) crossed with blue (*WP*) yields 25% white (*WW*), 50% blue (*WP*), and 25% purple (*PP*). (4) Purple (*PP*) crossed with blue (*PW*) produces no white petunias.

Alternate answer for #53.

(1) This cross is an example of incomplete dominance.

Key:

PP = purple
PW = blue
WW = white

Cross:

	P	W
W	PW	WW
W	PW	WW

Results:

50% white (WW)
50% blue (WP)

INCORRECT CHOICES:

(2) Key:

PP = purple

Cross:

	P	P
P	PP	PP
P	PP	PP

Results:

100% purple (PP)

(3) Key:

WP = blue

Cross:

	P	W
P	PP	PW
W	PW	WW

Results:

25% white (WW)
50% blue (WP)
25% purple (PP)

(4) Key:

PP = purple
PW = blue

Cross:

	P	P
P	PP	PP
W	PW	PW

Results:

50% purple (PP)
50% blue (PW)
0% white (WW)

54. **(4)** Grafting is a form of vegetative propagation in which new cells are produced by mitotic cell division and are, therefore, genetically like the parent cells.

55. **(5)** In situations involving incomplete dominance, the heterozygous individual (blue) does not resemble individuals with either of the homozygous combinations (black or white).

56. **(6)** Because black is dominant to its recessive allele, both pure black (BB) and hybrid black (Bb) guinea pigs are black.

57. **(1)** Sex-linked genes, such as the gene for color blindness, are located on the X chromosome.

58. **(3)** A change in a DNA molecule which can be inherited is known as a mutation. Radiation increases the mutation rates.

59. **(3)** When chromosomes fail to separate during disjunction, (nondisjunction), the result may be, as here, an extra chromosome in an in-

dividual. INCORRECT CHOICES: (1) DNA replication precedes both mitosis and meiosis and results in the production of two identical sets of genes to be separated in the ensuing nuclear division. (2) Cleavage refers to the first cell divisions of the animal embryo which produce the blastula. (4) Crossing-over is a type of chromosomal mutation in which portions of one chromatid are exchanged for corresponding portions of another during meiosis.

60. **(1)** Neither trait, as expressed in the question, appears dominant or recessive. There does appear to be a blending effect producing the intermediate oval. Hence, this is incomplete dominance. INCORRECT CHOICES: (2) Sex-linked traits deal with the transmission of traits which are recessive and linked to the X chromosome. (3) As neither trait appears dominant or recessive, this answer seems self-explaining. (4) This theory of Mendel's is based on the separation of gene pairs on a given pair of chromosomes which is completely independent of the separation of gene pairs on other chromosomes.

61. **(2)** Down's syndrome is due to an extra number 21 chromosome. INCORRECT CHOICES: (1) Color blindness is due to a change in a gene and not in chromosome number. (3) Hemophilia cannot be detected by observing the chromosomes since it is due to a single gene mutation. (4) Phenylketonuria is caused by a genetic mutation which prevents the synthesis of a particular enzyme.

62. **(3)** C represents the amniotic sac in which the embryo floats. INCORRECT CHOICES: (1) A depicts a follicle ripening in the ovary. (2) B is the placenta. (4) D represents the uterus.

63. **(4)** Working the problem out using the Punnett square method you would find that the probable genotypes are: AB, Ai, BB, Bi. The appearance of the O geneotype does not occur. INCORRECT CHOICES: (1), (2), and (3) All of these choices would appear in the Punnett square indicated above.

64. **(4)** Messenger RNA is single-stranded, contains uracil (U), and serves as a template for the bonding of transfer RNA molecules, each carrying its specific amino acid. INCORRECT CHOICES: (1) DNA is double-stranded and contains thymine (T) rather than uracil (U). (2) Protein consists of amino acids bonded together. (3) Transfer RNA molecules pick up and transfer amino acids from the cytoplasm to the messenger RNA molecule on the ribosomes.

65. **(3)** Protein synthesis, which occurs as amino acids are bonded in a sequence determined by the template messenger RNA, takes place on

the ribosomes. INCORRECT CHOICES: (1) DNA and RNA are synthesized in the nucleus. (2) Certain types of RNA are synthesized in the nucleolus. (4) The centrosome functions in the cell division of animal cells.

66. **(2)** Molecule 2 represents transfer RNA which functions to pick up and transfer amino acids from the cytoplasm to the ribosomes. INCORRECT CHOICES: (1) Enzymes are proteins made up of amino acids. (3) DNA serves as a template used in the replication of both itself and RNA. (4) DNA has the form of a double helix.

67. **(4)** This particular cross results in a ratio of 1:2:1. The results of the coin toss also approximate this ratio. INCORRECT CHOICES: (1) The results here would all be Aa or hybrid. (2) The results here would be a 1:1 ratio. (3) The results here would also be a 1:1 ratio.

68. **(2)** One of the stipulations of the Hardy-Weinberg principle is that the population of organisms studied reproduce sexually. INCORRECT CHOICES: (1) , (3) , and (4) All of these choices are presenting methods of asexual reproduction.

69. **(3)** Sexual reproduction has as one of its advantages the production of varieties through the following means: recombination of genes, as the question points out, and through mutations. INCORRECT CHOICES: (1) Homozygous parents have two identical genes for a trait; their offspring would show little variety. (2) According to Mendels Law of Independent Assortment, recessive characteristics do appear in future generations. (4) The theory of acquired characteristics as explained by Lamarck was disproved by Weismann.

70. **(1)** The complement of nitrogenous base pairs in DNA is simply this: adenine pairs with thymine, guanine pairs with cytosine. Accordingly, the proper sequence would be the one indicated. INCORRECT CHOICES: (2) Following the sequence outlined above, the base that is out of order here is the first A; it should be G. (3) Following that sequence, the U is misplaced and in fact does not appear in DNA at all but in RNA. (4) The base pair out of order is U which again is found only in RNA.

71. **(3)** Strand Z or m-RNA travel with their coded information to ribosomes to initiate protein production, that is with the aid of t-RNA. INCORRECT CHOICES: (1) The mitochondria are the powerhouses of the cell or, in other words, are involved in ATP formation. (2) Vacuoles, in unicellular organisms and plants, function in digestion and excretion. (4) Centrioles are involved in animal cell reproduction.

72. **(2)** The complementary bases in RNA are: adenine pairs with uracil, and guanine pairs with cytosine. Therefore, this choice is appropriate. INCORRECT CHOICES: (1) Thymine (T) is not found in RNA and therefore this choice is eliminated. (3) This sequence does not follow the sequence listed above. (4) Again we see the appearance of T which only appears in DNA.

73. **(1)** At this location in the DNA structure there are weak hydrogen bonds holding together the two strands of the ladder at the union between the nitrogen bases. They can be easily broken to release the double helix. INCORRECT CHOICES: (2) Number 2 is pointing to the sugar (deoxyribose) portion of the molecule. (3) Number 3 is pointing to the phosphate group. (4) Number 4 is pointing to a phosphate group.

74. **(4)** Down's syndrome results from the nondisjunction or failure of the number 21 chromosomes to separate. INCORRECT CHOICES: (1) As a result of crossing-over, corresponding portions of homologous chromosomes are exchanged. (2) Polyploidy produces cells with extra sets of chromosomes not one extra chromosome. (3) A gene mutation is a genetic not a chromosomal change.

75. **(4)** During mitosis, the genes on the chromosomes replicate to provide genetic information for new cells. INCORRECT CHOICES: (1) If a substance is easily destroyed by enzyme action, it is no longer a carrier of the same hereditary information. (2) All hereditary substances are not alike, resulting in variations for the same traits. (3) Not all cells have nuclei.

76. **(1)** Polypeptides refers to long chains of amino acids connected by peptide bonds. DNA transcribes m-RNA, which in turn travels to the ribosome and connects with t-RNA carrying specific amino acids, which are linked to form polypeptides. INCORRECT CHOICES: (2) Polysaccharides are long hydrocarbon units of glucose. (3) Lipids are composed of fatty acid molecules and glycerol. (4) The monosaccharides are simple sugars and include glucose, fructose, and galactose.

77. **(4)** It is seen that not only does the cell reproduce but that some organelles have this ability. The choroplasts and the mitochondria are two of these organelles. If this process occurs, then these organelles must contain genetic information enabling them to do this. INCORRECT CHOICES: (1) These organelles must contain genes for the process of reproduction to occur separately from the cell. (2) These organelles do not contain their own separate nuclei. (3) They contain their own genes but are controlled by genes in the nucleus of the cell.

78. **(3)** Individuals who are not able to taste PTC have the homozygous recessive genotype (p^2 = the homozygous recessive genotype. Therefore, $p^2 = 50 \div 200 = .25 = 25\%$. INCORRECT CHOICES: (1) , (2) , and (3) These choices do not logically appear in using the equation above.

79. **(2)** Using the Hardy-Weinberg principle, $2pq$ represents the hybrids or heterozygotes. $2pq$ appears on the graph in sections B and C. INCORRECT CHOICES: (1) A is homozygous and B is heterozygous. (3) C is heterozygous and D is homozygous. (4) A and D represent homozygous individuals.

80. **(1)** The p^2 and q^2 of the principle relate to the homozygous individuals. INCORRECT CHOICES: (2) A is homozygous and C is heterozygous. (3) B is heterozygous and D is homozygous. (4) B and C are both heterozygous.

81. **(3)** This graph depicts a Hardy-Weinberg population which has as one of its principles the factor of no isolation. We can see then, that any isolation would alter the stability. INCORRECT CHOICES: (1) According to the statement above, D might form a new species but it would not be the most immediate occurrence. (2) The mutation rate would not be higher in the isolated population. (4) No migrations would occur in isolated species.

82. **(2)** Chloroplasts have been shown to contain hereditary material. Since they can self-duplicate and are located in the cytoplasm, this is called cytoplasmic inheritance. INCORRECT CHOICES: (1) Chromosomal nondisjunction affects the numbers of chromosomes in the offspring. (3) Bacterial transformation is the replacement of dissolved DNA from one cell to another. (4) Gene mutation causes changes in the genotype and phenotype of the offspring.

83. **(2)** The changed base sequence is incorporated into the genotype and will be passed on when DNA replicates. INCORRECT CHOICES: (1) The DNA pictured occurs only on one chromosome. (3) The phenotype deals with the physical appearance. The change illustrated deals with the genotype. (4) Since the coded sequence has been changed, the genotype of the offspring will most definitely be changed.

84. **(1)** Chromosomes are composed of genes and genes are composed of DNA. Therefore, a change in DNA that would cause a mutation would mutate the gene. INCORRECT CHOICES: (2) Nucleic acid replication would be the production of identical DNA molecules. (3) Protein synthesis involves the m-RNA traveling to the ribosome and connecting with amino acid units carried by t-RNA. (4) Gene replication would be the duplication of identical genes.

85. **(3)** Sickle-cell anemia is caused by the substitution of the amino acid valine for the amino acid glutamic acid. This substitution has its basis in the mutation of the genetic code as pictured above. The change in the sequence of bases will cause the substitution of amino acid molecules at the ribosome. INCORRECT CHOICES: (1) Down's syndrome results from an extra chromosome on pair 21 which occurs in all body cells. (2) Polyploidy is the condition in which the cells contain more than twice the monoploid number of chromosomes. (4) Phagocytosis is the engulfment of food particles which are too large to pass through the cell membrane.

86. **(1)** DNA replication must precede the division of the chromosomes in mitosis. INCORRECT CHOICES: (2) Cyclosis is the movement or circulation of the cytoplasm. (3) Pinocytosis occurs when the cell membrane forms a pocket to take in large molecules. (4) Protein synthesis requires reading the DNA code, not the replication of the DNA molecule.

87. **(1)** Mitochondria and chloroplasts contain small amounts of their own DNA and are thus capable of duplicating themselves. INCORRECT CHOICES: (2) Cell walls are built by plant cells and provide support and protection for them. (3) As above, cell walls are made by the cells within them. (4) Vacuoles are membrane-surrounded spaces filled with water and dissolved materials.

88. **(4)** Environmental conditions affect the phenotypes of the organisms. It has also been shown, in particular with the peppered moths, that environmental changes also affect the genotypes. INCORRECT CHOICES: (1) Identical twins, as the term implies, would have the same genotypes. (2) Identical twins develop from one fertilized egg. (3) Identical twins are born minutes apart and this time element has no bearing on their chromosomal makeup.

89. **(2)** Chloroplasts are structures in the cytoplasm which can duplicate themselves. INCORRECT CHOICES: (1) Enzymes cannot replicate: they are manufactured under the direction of DNA. (3) and (4) The principles of dominance and segregation apply to gene not chloroplast reproduction.

90. **(4)** Independent assortment refers to the independent inheritance of two traits; here, hair color and eye color are inherited independently of each other. INCORRECT CHOICES: (1) Multiple alleles refers to situations in which more than two alleles are involved in the inheritance of a single trait. (2) Incomplete dominance describes a situation in which neither allele in a heterozygous individual is dominant. therefore, heterozygous individuals do not resemble individuals with either of the homozygous

combinations. (3) Sex linkage refers to the inheritance of traits, the genes of which are located on the X chromosome.

91. **(1)** Attraction is weakest at hydrogen bond sites. It is a light atom and has a weak chemical bonding power. INCORRECT CHOICES: (2) Carbon atoms are strongly bonded to each other and to other atoms in DNA. (3) Phosphate groups are more strongly bonded to sugar molecules in DNA. (4) Ribose groups are not part of a DNA molecule.

92. **(2)** DNA molecules provide the code for the kind of sequence of amino acids in the synthesis of proteins. INCORRECT CHOICES: (1) Enzymes catalyze the formation of peptide bonds. (3) DNA molecules are not found in the cytoplasm. (4) ATP supplies energy for protein synthesis.

93. **(2)** The diagram presents a somewhat complex picture of DNA replication and transcription of RNA. The complementary bases for the bases shown in number 7 would those of this choice. INCORRECT CHOICES: (1) Thymine is substituted for the correct answer of uracil. In addition thymine is not a base in RNA. (3) This choice is the identical sequence for that in number 7. (4) This choice shows the nitrogen bases completely misaligned.

94. **(4)** The letters represent the purine and pyrimidine groups that composes the nitrogen bases. INCORRECT CHOICES: (1) and (3) The phosphates and sugars make up the sides, not the rungs, of the DNA and RNA molecules. (2) The nucleic acid is the entire structure which is composed of sugar, phosphate, and nitrogen bases.

95. **(1)** According to the process of protein synthesis, the m-RNA leaves the nucleoplasm transcribed with a specific set of codes from DNA, and travels to the ribosome to initiate protein synthesis. INCORRECT CHOICES: (2) The mitochondria are associated with the production of ATP. (3) The nucleolus aids in the synthesis of protein. (4) The centrosome plays a role in the mitotic division of animal cells.

96. **(2)** In looking at the keyed diagram or the pedigree, we see that individual marked 5 is a half-shaded circle. In looking at the key below the pedigree, we find this symbol to mean this choice. The genotype is arrived at by saying the individual is a carrier female, which again is $X^N X^n$. INCORRECT CHOICES: (1) The genotype given is for a woman who has normal vision. According to the keyed diagram the normal woman would be an open circle. (3) The genotype given is for an afflicted woman. The key in the diagram that applies would be the shaded circle. This of course is not the case. (4) This genotype is for an afflicted male. If this

were the case in number 5 the circle would be replaced by a shaded box.

97. **(4)** The choice of 12 and 13 would seem logical if we were to look at the pedigree. Both are normal females who have the same parents. INCORRECT CHOICES: (1) Five and six could not possibly be identical twins in that their genotypes are different. (2) Seven and eight are husband and wife, not identical twins. (3) Nine and eleven again have different genotypes and could not be identical.

98. **(4)** Working out the Punnett square for this cross we would have as the P_1 generation the following: $X^n Y$ crossed with $X^N Y^N$; this would produce 100% normal vision offspring. INCORRECT CHOICES: (1), (2), and (3) These choices do not appear feasible according to the results obtained from the Punnett square.

99. **(1)** According to the Punnett square for this mating: P_1 is $X^N Y$ crossed with $X^n X^n$; the choice of number 1 is logical. INCORRECT CHOICES: (2), (3), and (4) These choices do not appear feasible according to the results obtained from the Punnett square.

100. **(3)** The diagram is showing us a colorblind male who needs only one gene associated with the X chromosome to be colorblind and the female who is a carrier and possesses only one gene for colorblindness on one X chromosome. INCORRECT CHOICES: (1) The female number 5 is not colorblind. (2) Only the female has normal vision, the male is colorblind. (4) The male can have only one gene for colorblindness.

6

Evolution and Diversity

A. Evidence for Evolution
   1. The fossil record
   2. Comparative anatomy
   3. Comparative embryology
   4. Comparative biochemistry
   5. Geographic distribution

B. Theories of Evolution
   1. Lamarck's theory
   2. Darwin's theory
   3. Modern theory
      a. Production of variations
      b. Population genetics
         (1) Population
         (2) Gene pool
      c. Transmission of variations
         (1) Factors influencing transmission
            (a) The environment
            (b) Isolation
         (2) Patterns of change
            (a) Change within a species
            (b) Development of new species

C. A Modern Classification System

D. Origin and Early Evolution of Life
   1. The heterotroph hypothesis
      a. Assumptions
         (1) Raw materials
            (a) Inorganic compounds
            (b) Energy sources
         (2) Synthesis
         (3) Nutrition
         (4) Reproduction

2. Heterotroph to autotroph
3. Anaerobe to aerobe

The following questions on Evolution and Diversity have appeared on previous Regents Examinations.

Directions: Each question is followed by four choices. Underline the correct choice.

1. According to the heterotroph hypothesis, heterotrophs changed the early atmosphere by adding
   (1) methane
   (2) oxygen
   (3) ammonia
   (4) carbon dioxide

2. Which condition will cause instability in a gene pool?
   (1) small populations
   (2) random mating
   (3) no migration
   (4) lack of mutations

3. A genetic basis for evolution is provided by
   (1) Lamarck's thoery
   (2) Darwin's theory
   (3) modern evolutionary theory
   (4) the fossil record

4. The following genotypes occurred in a stable, randomly mating population:
   $$AA—16\%$$
   $$Aa—48\%$$
   $$aa—36\%$$

   What would be the expected percentage of hybrids in the next generation?
   (1) 16%
   (2) 24%
   (3) 48%
   (4) 96%

5. Which best explains the abundance of marsupials in Australia?
   (1) migration of reptiles to that continent
   (2) geographic isolation
   (3) the failure of placental mammals to survive in Australia
   (4) the extinction of the common ancestor of the marsupials and the placental animals

Directions: Base your answers to questions 6 through 8 on your knowledge of the Hardy-Weinberg principle and on the information below representing the frequency of alleles for eye color in a population.

Allele	Frequency
B (brown eyes)	0.7
b (blue eyes)	0.3

6. According to the Hardy-Weinberg principle, which term would represent the frequency for hybrid brown-eyed individuals?
   (1) $2pq$ (3) p^2
   (2) $2p$ (4) q^2

7. What is the frequency of homozygous brown-eyed individuals in this population?
   (1) 0.09   (3) 0.49
   (2) 0.21   (4) 0.96

8. What is the frequency for recessive individuals in this population?
   (1) 0.09   (3) 0.49
   (2) 0.21   (4) 0.96

9. In modern classification, protozoa, algae, fungi, slime molds, and bacteria are known as
   (1) bryophytes   (3) protists
   (2) plants   (4) animals

10. Based on Lamarck's theory, what is an explanation for the evolution of a tailless species of monkey from an ancestor that had a tail?
    (1) The gene for the tailless trait is dominant over the gene for the tailed trait.
    (2) A mutation occurred in the tailed species.
    (3) The tailless monkey did not overpopulate the inhabited area.
    (4) The tail was no longer needed to help the monkey escape from its predators.

11. The theory that birds and reptiles share common ancestry is supported by the evidence that they both
    (1) occupy similar niches
    (2) have similar environmental requirements
    (3) show structural similarities during their development
    (4) have evolved as separate groups at about the same time

12. Lamarck proposed that new organs evolved according to the
    - (1) needs of the organism
    - (2) process of natural selection
    - (3) role of mutation
    - (4) sorting out of genes

13. The Australian continent contains many species of marsupials found nowhere else in the world. According to modern evolutionary theory, the existence of these species is most likely the result of
    - (1) geographic isolation
    - (2) genetic stability
    - (3) no mutations
    - (4) random mating

14. Which is *not* a part of Darwin's theory of evolution?
    - (1) natural selection
    - (2) struggle for existence
    - (3) variation due to mutations
    - (4) overproduction of organisms

15. In the equation $p^2 + 2pq + q^2 = 1$ from the Hardy-Weinberg principle, p represents the frequency of the
    - (1) dominant allele
    - (2) recessive allele
    - (3) homozygous individuals
    - (4) heterozygous individuals

16. Which evidence of evolution is illustrated when scientists study the similarities of nucleic acids in vertebrates?
    - (1) comparative anatomy
    - (2) comparative biochemistry
    - (3) comparative embryology
    - (4) genetic isolation

17. The extent to which a given gene occurs in a population is known as the gene
    - (1) number
    - (2) action
    - (3) pool
    - (4) frequency

18. According to modern evolutionary theory, which factor would probably contribute most to the evolution of a new species?
    - (1) stable environmental conditions
    - (2) low mutation rate
    - (3) isolation of part of a population
    - (4) a stable gene pool

19. Of the following factors, which has the *least* effect on the rate of evolution of a species?
    - (1) variation
    - (2) natural selection
    - (3) asexual reproduction
    - (4) isolation

20. Which diagram on the next page best illustrates a relationship between Kingdom, Phylum, and Genus?

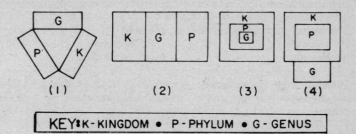

KEY: K - KINGDOM • P - PHYLUM • G - GENUS

21. The neck of a giraffe and the neck of a human both have 7 vertebrae. This is considered evidence for evolution from the study of
    (1) the fossil record
    (2) comparative embryology
    (3) comparative anatomy
    (4) geographical distribution

22. In undisturbed layers of the Earth's crust, the bones of dinosaurs were found in a lower layer than the bones of mammals. This would lead us to assume that dinosaurs
    (1) lived in caves
    (2) were burrowing animals
    (3) appeared earlier than mammals
    (4) appeared later than amphibians

23. In any given location, the term "population" refers to all of the
    (1) different animal species
    (2) different plant species
    (3) plant and animal species
    (4) members of a species

24. The occurrence of antibiotic-resistant bacteria and DDT-resistant house-flies results from
    (1) the inheritance of acquired characteristics by the offspring
    (2) the use and disuse of resistant traits by parents and offspring
    (3) genetic changes which enabled these organisms to adapt to new environmental conditions
    (4) the need of flies and bacteria to become immune to pesticides and antibiotics

25. If members of the same species have been geographically isolated from each other for an extended period of time, which will they most likely exhibit?
    (1) mutations identical to each other
    (2) random recombination occurring in the same manner
    (3) evolution of traits of high adaptive value for their particular environments
    (4) evolution into two new species which will have no problem interbreeding

26. "The human earlobe, because it is not used, will probably disappear from the human population in the future."
This statement reflects a theory proposed by
(1) Darwin   (3) Mendel
(2) Watson   (4) Lamarck

27. Which term describes appendages that may have different functions, but are similar in structure and are assumed to have the same evolutionary origin?
(1) fossils   (3) homologous
(2) homozygous   (4) mutations

28. According to the Hardy-Weinberg principle, which condition would favor stability of a gene pool from generation to generation?
(1) selective mating among individuals in a population
(2) a high rate of mutation in a population
(3) a population comprised of a small number of individuals
(4) no migration of individuals into or out of the population

29. A biological classification system is most useful for
(1) understanding relationships between different types of organisms
(2) determining the age of fossils found in different rock layers
(3) proving the theory of evolution developed by Darwin
(4) proving the heterotroph hypothesis

30. An important part of Lamarck's theory of evolution was the concept of
(1) homologous structures
(2) mutation
(3) struggle for existence
(4) the inheritance of acquired characteristics

31. The total of all the heritable genes for all the traits of a population constitutes the
(1) genetic distribution   (3) gene pool
(2) dominant genes   (4) recessive genes

32. In the process of evolution, the effect of the environment is to
(1) prevent the occurrence of mutations
(2) act as a selective force on variations in species
(3) provide conditions favorable for the formation of fossils
(4) provide stable conditions favorable to the survival of all species

33. One factor that Darwin was unable to explain satisfactorily in his theory of evolution was
    (1) natural selection
    (2) overproduction
    (3) survival of the fittest
    (4) the source of variations

34. Insulin extracted from cattle is so similar to the insulin of humans that it is used to treat human diabetics. This similarity is an example of a type of evidence for evolution gained from comparative
    (1) anatomy
    (2) embryology
    (3) biochemistry
    (4) fossil study

35. Certain insects resemble the twigs of trees on which they live. The most probable explanation for this resemblance is that
    (1) the trees caused a mutation to occur
    (2) no mutations have taken place
    (3) natural selection has favored this trait
    (4) the insects needed to camouflage themselves

36. The total of all the heritable genes for all the traits in a given population is called a gene
    (1) pool
    (2) mutation
    (3) action
    (4) frequency

37. In a class of 100 students, 25 were found to be nontasters of PTC paper. The ability to taste is a dominant trait. According to the Hardy-Weinberg formula, $p^2 + 2pq + q^2 = 1$, the probable frequency of homozygous tasters (TT) is
    (1) 10%
    (2) 25%
    (3) 50%
    (4) 75%

38. Variations among offspring are most frequently produced by the combined effects of
    (1) vegetative reproduction and chromosome mutation
    (2) gene mutations and sexual reproduction
    (3) binary fission and sexual reproduction
    (4) gene mutations and parthenogenesis

Directions: Base your answers to questions 39 and 40 on the information below and on your knowledge of biology.

Within a particular species of meadowmouse, two variations of coat color normally exist. brown Coat (*B*) is dominant over white coat (b). The frequency of the dominant allele equals 0.5 and the frequency of the recessive allele equals 0.5.

39. According to the Hardy-Weinberg equation, what would be the frequency for the heterozygous genotype?
   (1) 25%   (3) 75%
   (2) 50%   (4) 100%

40. If brown mice are selected for mating, what will be the future effect on the meadowmouse population?
   (1) The value of $2pq$ will remain unchanged.
   (2) p and q will remain unchanged.
   (3) p^2 will increase as q^2 decreases.
   (4) q^2 will increase as p^2 decreases.

41. A scientist studying fossils in undisturbed layers of rock identified a species that, he concluded, changed little over the years. Which observation probably would have led him to this conclusion?
   (1) The simplest fossil organisms appeared only in the oldest rocks.
   (2) The simplest fossil organisms appeared only in the newest rocks.
   (3) The same kind of fossil organisms appeared in old and new rocks.
   (4) No fossil organisms of any kind appeared in the newest rocks.

Directions: Base your answers to questions 42 through 44 on your knowledge of biology and on the following chart, which shows data from a population genetics survey of a certain species.

	1900	1910	1920	1930	1940	1950	1960	1970
Frequency of Allele *B*	0.99	0.81	0.64	0.49	0.36	0.25	0.16	0.10
Frequency of Allele *b*	0.01	0.19	0.36	0.51	0.64	0.75	0.84	0.90

42. Which is the best interpretation of the information in the chart relative to the frequencies of alleles B and b?
    (1) B increased as b decreased.
    (2) B decreased as b increased.
    (3) Both B and b are decreasing in frequency.
    (4) Both B and b are increasing in frequency.

43. The best explanation for the data in the chart is that
    (1) allele b was needed by the species and therefore appeared
    (2) organisms carrying allele b had a reduced rate of survival
    (3) a change in environment favored the trait controlled by allele B
    (4) a change in environment favored the trait controlled by allele b

44. In 1970, what was the frequency of homozygous (BB) individuals?
    (1) 1%   (3) 81%
    (2) 18%   (4) 100%

45. In animals, the largest number of different organisms are found in the classification group known as
    (1) kingdom   (3) genus
    (2) phylum   (4) species

46. Which is classified as a protist?
    (1) sponge   (3) Hydra
    (2) Paramecium   (4) moss

47. Most modern biologists agree that an ideal classification system should reflect
    (1) nutritional similarities among organisms
    (2) habitat requirements of like groups
    (3) distinctions between organisms based on size
    (4) evolutionary relationships among species

48. Darwin observed that different, but closely related, species of finches filled the diverse environmental niches on the different Galapagos Islands. The filling of these environmental niches is known as
    (1) acquired characteristics   (3) common ancestry
    (2) blending inheritance   (4) adaptive radiation

49. In one modern classification system, organisms are grouped into three
    (1) kingdoms   (3) genera
    (2) phyla   (4) species

50. Some species of algae are unicellular, motile, and photosynthetic. Under one modern system of classification, they would be classified as
    (1) mollusks
    (2) fungi
    (3) protists
    (4) tracheophytes

51. Which is one basic assumption of the heterotroph hypothesis?
    (1) More complex organisms appeared before less complex organisms.
    (2) Living organisms did not appear until there was oxygen in the atmosphere.
    (3) Large autotrophic organisms appeared before small photosynthesizing organisms.
    (4) Autotrophic activity added molecular oxygen to the environment.

52. According to the heterotroph hypothesis, the most primitive organism was an
    (1) aerobic heterotroph
    (2) anaerobic heterotroph
    (3) aerobic autotroph
    (4) anaerobic autotroph

53. According to the modern theories of evolution, which of the following factors would be least effective in bringing about species changes?
    (1) geographic isolation
    (2) changing environments
    (3) genetic recombination
    (4) asexual reproduction

54. Application of Mendelian principles to the study of gene behavior in populations is known as
    (1) adaptive radiation
    (2) population genetics
    (3) gene frequency
    (4) variations

55. In a given gene pool, the frequency of the dominant gene is 0.5. Which is true of this gene pool?
    (1) More organisms will show the dominant than the recessive characteristic.
    (2) More organisms will show the recessive than the dominant characteristic.
    (3) Equal numbers of organisms will show the dominant and the recessive characteristics.
    (4) There will be equal numbers of hybrids and recessives.

56. Modern evolutionary theory has modified the theory of natural selection by
    (1) considering survival of the fittest to be invalid
    (2) showing that competition does not exist within a species
    (3) including a genetic basis for change and variation
    (4) accepting the theory of use and disuse

57. Two organisms can be considered to be of different species if they
    (1) cannot mate with each other and produce fertile offspring
    (2) live in two different geographical areas
    (3) mutate at different rates depending on their environment
    (4) have genes drawn from the same gene pool

58. What would be the most probable effect of reproductive isolation in a population?
    (1) It has no effect on variations in the species.
    (2) It favors the production of new species.
    (3) It prevents the occurrence of mutations.
    (4) It encourages the mixing of gene pools.

59. The comparative sciences of anatomy, embryology, and biochemistry provide evidences of evolution which support the concept of
    (1) genetic dominance   (3) geographic isolation
    (2) fossil formation   (4) common ancestry

60. In a certain area, DDT-resistant mosquitoes now exist in greater numbers than ten years ago. What is the most probable explanation for this increase in numbers?
    (1) Genetic differences permitted some mosquitoes to survive DDT use.
    (2) Mosquito eggs were most likely to have been fertilized when exposed to DDT.
    (3) DDT acted as a reproductive hormone for previous generations of mosquitoes.
    (4) DDT serves as a new source of nutrition.

61. According to the Hardy-Weinberg principle, under which conditions would a population's gene pool tend to remain constant?
    (1) large size and migration
    (2) large size and random mating
    (3) small size and frequent mutations
    (4) small size and nonrandom mating

62. The graph below represents the percent of variation for a given trait in four different populations of the same species. These populations are of equal size and inhabit similar environments.

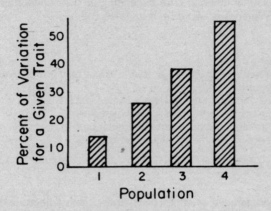

In which population is the greatest number of individuals most likely to survive significant environmental changes related to this trait?
(1) 1
(2) 2
(3) 3
(4) 4

63. Which factor has the greatest effect on the rate of evolution of animals?
(1) environmental changes
(2) use and disuse
(3) asexual reproduction
(4) vegetative propagation

64. In an attempt to explain the diversity of living things, Darwin's theory of natural selection
(1) proved evolution took place
(2) described how mutations produced variations
(3) showed that only the largest animals survive
(4) described how evolution could have occurred

65. Occasionally hospitals have outbreaks of staphylococcus bacteria infections which resist treatment with antibiotics that were once effective. What is the most acceptable explanation for this condition?
(1) Due to selective forces, new staphylococcus strains have become more abundant.
(2) Newer antibiotics are weaker than those used in the past.
(3) Laboratory strains of staphylococcus have been weakened.
(4) Patients have become less susceptible to bacterial infection.

Directions (66–69): For each phrase in questions 66 through 69, select the evolutionary term, *chosen from the list below,* that is best described in that phrase.

Terms Related to Evolution

(1) Adaptive radiation
(2) Gene pool
(3) Natural selection
(4) Reproductive isolation
(5) Comparative embryology

66. Total of all the inherited DNA for all traits in a given species in a particular location

67. The tendency of species members to fill the available environmental niches

68. An increase in the number of mosquitoes resistant to DDT

69. The resemblance of mollusk larvae to larvae of annelids

70. In the equation, $p^2 + 2pq + q^2 = 1$, which provides geneticists with a method for determining the distribution of genes in a population, $2pq$ represents the frequency of
(1) homozygous dominant individuals
(2) homozygous recessive individuals
(3) heterozygous individuals
(4) dominant and recessive alleles

71. Which biological process is *not* normally a basis for the production of variations within a species?
(1) gene mutations   (3) binary fission
(2) recombination of genes (4) crossing-over

72. In guinea pigs, the gene for black fur *(B)* is dominant over the gene for white fur *(b)*. If 16% of a guinea pig population has white fur, what is the percentage of homozygous dominant individuals?
(1) 4%   (3) 16%
(2) 6%   (4) 36%

Answers

1. 4	19. 3	37. 2	55. 1
2. 1	20. 3	38. 2	56. 3
3. 3	21. 3	39. 2	57. 1
4. 3	22. 3	40. 3	58. 2
5. 2	23. 4	41. 3	59. 4
6. 1	24. 3	42. 2	60. 1
7. 3	25. 3	43. 4	61. 2
8. 1	26. 4	44. 1	62. 4
9. 3	27. 3	45. 1	63. 1
10. 4	28. 4	46. 2	64. 4
11. 3	29. 1	47. 4	65. 1
12. 1	30. 4	48. 4	66. 2
13. 1	31. 3	49. 1	67. 1
14. 3	32. 2	50. 3	68. 3
15. 1	33. 4	51. 4	69. 5
16. 2	34. 3	52. 2	70. 3
17. 4	35. 3	53. 4	71. 3
18. 3	36. 1	54. 2	72. 4

Explanatory Answers

1. **(4)** The first heterotrophs may have added carbon dioxide to the early atmosphere as a result of their anaerobic respiration. INCORRECT CHOICES: (1) and (3) Methane and ammonia are believed to have been part of the early atmosphere. (2) Oxygen is believed to have been added to the atmosphere by the action of autotrophs in photosynthesis.

2. **(1)** In small populations chance events may alter the gene frequencies. INCORRECT CHOICES: (2) Random mating leads to a stable gene pool. (3) If there is no movement of individuals in or out of the population, gene

frequencies will tend to remain the same. (4) Mutation involves a change in the genes and therefore, a change in the gene frequencies.

3. **(3)** Modern evolutionary theory provides a genetic basis for evolution. INCORRECT CHOICES: (1) Lamarck was in error in suggesting that useful characteristics acquired in one generation could be transmitted to the next. (2) Darwin's theory of natural selection did not provide a genetic basis for variations within a species. (4) The fossil record supplies evidence supporting the concept of evolution but does not suggest how evolution occurs.

4. **(3)** In a stable, randomly mating population, gene frequencies and genotype frequencies stay the same from generation to generation. INCORRECT CHOICES: (1) The percentage of homozygous dominant individuals in the next generation would be 16%. (2) and (4) In a stable population, the percentage of hybrids (*Aa*) will neither decrease nor increase.

5. **(2)** Geographic isolation of marsupials in Australia enabled them to survive without competition from placental mammals. INCORRECT CHOICES: (1) Migration of reptiles does not explain the abundance of marsupials. (3) Placental mammals introduced into Australia have done very well there. (4) The extinction of the common ancestor does not explain the geographic distribution of placental and marsupial mammals.

6. **(1)** Hybrid brown-eyed individuals are heterozygous and have one allele for brown (*p*) and another for blue (*q*) with a total frequency of $2pq$ (0.42) in the population. INCORRECT CHOICES: (2) $2p$ is not a term in the Hardy-Weinberg equation. (3) p^2 represents the frequency for the homozygous dominant trait. (4) q^2 represents the frequency of the homozygous recessive trait.

7. **(3)** The frequency of homozygous brown-eyed individuals is equal to 0.7^2, or p^2, or 0.49. INCORRECT CHOICES: (1) 0.09 represents the frequency of the blue trait, 0.3^2 or q^2. (2) 0.21 represents half the frequency of the heterozygous or hybrid individuals, ½ of $2pq$, or 0.07×0.03. (4) 0.96 does not represent any of the gene frequencies in this population.

8. **(1)** 0.09 represents the frequency of the blue trait, 0.03^2 or q^2. INCORRECT CHOICES: (2) 0.21 represents half the frequency of the heterozygous or hybrid individuals, ½ of $2pq$, or 0.07×0.03. (3) 0.49 represents the frequency of the homozygous brown-eyed individuals, 0.07^2, or p^2. (4) 0.96 does not represent any of the gene frequencies in this population.

9. **(3)** Protists are one-celled organisms. All these are single celled. INCORRECT CHOICES: (1), (2), and (4) Bryophytes, plants, and animals are multicelled organisms. A bryophyte is a plant.

10. **(4)** Lamarck put forth the theory of use and disuse. If an organ or structure was not used or needed, he believed that structure was not passed on as a trait to the next generation. INCORRECT CHOICES: (1) Darwin's law of dominance was not part of Lamarck's theory. (2) Mutations were not known as scientific occurrences when Lamarck proposed his theories. (3) The principle of "overpopulation" is attributed to Darwin.

11. **(3)** The similarities shown by vertebrate embryos during development is evidence of common ancestry. INCORRECT CHOICES: (1) An ecological niche describes the role that a species plays in the community. (2) Organisms that have similar environmental requirements occupy the same habitat. (4) Evidence shows that reptiles existed before birds.

12. **(1)** Lamarck's theory assigned purpose to evolution and postulated that organs arise according to the needs of the organism. INCORRECT CHOICES: (2) Darwin, not Lamarck, described the process of natural selection. (3) When Lamarck proposed his theory the role of mutation in evolution was not understood. (4) Genes had not been discovered when Lamarck postulated his theory.

13. **(1)** Due to the theory of continental drift, we assume that the species in Australia were isolated from interbreeding units elsewhere. According to the isolation idea these organisms in Australia adapted to and changed to meet their environment without regard to the path followed by their relatives elsewhere. INCORRECT CHOICES: (2) Genetic stability is not one characteristic of an isolated species. (3) Mutations had to occur in that a very few number of genes in the gene pool were available. (4) Random mating would not account for the variations but rather the stability of the gene pool.

14. **(3)** Variations due to mutations could not have been one of Darwin's ideas because the idea of mutations was not developed until much later. INCORRECT CHOICES: (1), (2), and (4) All of these choices are part of Darwin's theory.

15. **(1)** p represents the dominant individual in the population. INCORRECT CHOICES: (2) The recessive allele is represented by q. (3) The homozygous individuals are represented by p^2 and q^2. (4) The heterozygous individuals are represented by $2pq$.

16. **(2)** Since nucleic acids are organic compounds and the study of these compounds in biology is biochemistry, then it would stand to reason that the comparison of them would be comparative biochemistry. INCORRECT CHOICES: (1) Comparative anatomy studies the structures of organisms. (3) Comparative embryology studies the embryos of organisms. (4) Genetic isolation refers to geographical removal of one gene pool from the others.

17. **(4)** The frequency is the rate at which something occurs. The occurrence of a particular gene in a population then would correctly be called the frequency. INCORRECT CHOICES: (1) The gene number might more appropriately be applied to a gene map showing location and structure. (2) Gene action would logically refer to the activity of the gene. (3) A gene pool is all of the genes of a population living in one area.

18. **(3)** Isolating a gene pool would cause a greater evolutionary rate in that no outside forces such as interbreeding could occur. Therefore the isolated species would evolve in a different direction than the nonisolated species. INCORRECT CHOICES: (1), (2), and (4) These choices refer to the Hardy-Weinberg principle for the stable populations in which few mutations would occur.

19. **(3)** Asexual reproduction produces organisms which are identical in genetic make-up and provide little variation for evolution. INCORRECT CHOICES: (1) Variation would have a great effect on the rate of evolution because some individuals would be better suited to their environments than others; they would reproduce and pass on their desirable variations. (2) Natural selection states that those best adapted survive and reproduce; as time and generations continue, evolution occurs. (4) Isolation speeds up the process of evolution or change because keeping variations within separate groups causes the gene frequencies to shift.

20. **(3)** The picture shows the genus as the smallest part of the total, the Phylum the next largest, and the Kingdom encompassing the other two. INCORRECT CHOICES: (1) The three categories are pictured as being in equal amounts whereas the genus is the smallest in number followed by the Phylum and Kingdom. (2) Here the order of occurrence is wrong. (4) In this diagram, the Genus is pictured as being outside the relationship of the other two.

21. **(3)** Comparative anatomy involves study of similarities in structures of different organisms; such similarities suggest evolution from a common ancestral form. INCORRECT CHOICES: (1) The fossil record consists of the preserved remains in the rocks of organisms which lived in the past. (2)

Comparative embryology involves study of embryonic structures in different organisms. (4) Geographical distribution refers to the ranges of organisms over the earth.

22. **(3)** It is assumed that the lowest layers are the oldest ones, and therefore, that bones found in the lower layers belonged to animals that lived before the animals whose bones are found in higher layers. INCORRECT CHOICES: (1) The presence of cavelike formations in the rock layers in which the fossils were found might be evidence that animals lived in caves. (2) Evidence of burrowing might be found in the rock layers in which the fossils were found. (4) No mention of amphibian fossils is made here.

23. **(4)** A population consists of all the members of one species inhabiting a given geographic location. INCORRECT CHOICES: (1) The different animal species in a given location consist of many separate populations. (2) The different plant species in a given location consist of many separate populations. (3) All of the plant and animal species in a given location comprise the biological community for that location.

24. **(3)** The addition of antibiotics and DDT to the environments of bacteria and flies increased the survival value of inherited resistance. INCORRECT CHOICES: (1) Acquired characteristics cannot be inherited. (2) A trait is not inherited merely because it is useful. (4) There is no evidence for assigning purpose to evolutionary change.

25. **(3)** The resulting individuals will have traits which have given them a higher survival rate in each area. INCORRECT CHOICES: (1) The probability of exact genetic changes due to mutations is highly unlikely. (2) Random recombination over an extended time would be a rare occurrence. (4) Different species when interbred result in sterile offspring.

26. **(4)** Lamarck's theory of use and disuse suggested that an organ which is not used will decrease in size. INCORRECT CHOICES: (1) Darwin's theory of natural selection suggested that the environment selected those individuals whose adaptations best fitted them for survival. (2) Watson's contributions to biological thought involved the molecular nature of the gene. (3) Mendel developed principles which form the basis for our understanding of the principles of heredity.

27. **(3)** Homologous structures are similar in form and evolutionary origin but differ in function. INCORRECT CHOICES: (1) Fossils are remains from organisms that lived in the past. (2) Homozygous refers to organisms that have two of the same genes for a trait. (4) Mutations are genetic changes that are inherited by subsequent generations.

28. **(4)** The frequency of specific genes in the gene pool would change if individuals left or entered a selected population. INCORRECT CHOICES: (1) Selective mating would force an increase in the frequencies of certain genes and a decrease in others. (2) Mutations would introduce new genes into the gene pool resulting in a change in the gene frequencies in an unstable population. (3) The possibility of changing the frequency of a gene or trait is greater in a small population where the chances are greater that recombination will occur.

29. **(1)** A useful biological classification system should reflect evolutionary similarities and differences making it possible to understand relationships between organisms. INCORRECT CHOICES: (2) The age of fossils may sometimes be determined by various methods, including the use of radioactive "clocks." (3) Evidence supporting Darwin's theory of evolution comes from many studies including comparative anatomy and embryology. (4) The heterotroph hypothesis is an explanation for the origin of life.

30. **(4)** Lamarck believed that useful characteristics acquired in one generation may be transmitted to the next. INCORRECT CHOICES: (1) Homologous structures are structures in different organisms with striking similarities in anatomical features. (2) The idea of mutation as a source of variation within a species is part of modern evolutionary theory. (3) Darwin conceived of a struggle for existence among the individuals in a population leading to a natural selection of those individuals with the most favorable adaptations.

31. **(3)** The gene pool for a population consists of all the genes for all the inheritable traits for that population. INCORRECT CHOICES: (1) The ratio of, or frequency with which, inheritable traits are found in the population is the genetic distribution. (2) Dominant genes are responsible for those traits of alleles which are displayed in the hybrid or heterozygous condition. (4) Recessive genes are responsible for those traits of alleles which are not displayed in the hybrid or heterozygous condition.

32. **(2)** Species which survive are those which have best adapted to the forces in the environment. INCORRECT CHOICES: (1) Mutations may increase due to a specific type of environment, e.g., uranium in an area. (3) The forces of nature in the environment may destroy fossil remains, e.g., rain on imprints. (4) In the process of evolution or change, some species will have traits more favorable to their survival than others, even if a stable environment were possible.

33. **(4)** Variations or mutations were not realized until Mendel's ideas were accepted years later. Then, the idea of mutations wasn't given any factual basis until still later. Darwin was expressing mere physical observation

in discussing variations. INCORRECT CHOICES: (1), (2), (3) These ideas of Darwin were well substantiated through physical evidence and the writings of others such as Malthus.

34. **(3)** Insulin is a chemical produced in the islets of Langerhans, a specialized area of pancreatic cells. To produce similar insulin, two species of organisms would necessarily have to have had a common ancestor to account for the closeness of their ability to synthesize similar insulin. INCORRECT CHOICES: (1) Anatomy of the two species would have no bearing on their biochemical abilities. (2) The embryos of mammals would all appear similar at some point and would not determine their biochemical synthesizing abilities. (4) Fossil study would merely tell us about similarities in structure.

35. **(3)** Insects that resemble their environment are better adapted to avoid predators. INCORRECT CHOICES: (1) Mutations occur spontaneously. (2) Mutations must have taken place in order for the insects to develop this form of camouflage. (4) Mutations occur at random and not according to need.

36. **(1)** The gene pool is the sum of all the genes collectively present in a population. INCORRECT CHOICES: (2) A mutation is a change in a gene that is heritable. (3) Genes act to code hereditary traits. (4) The percentage of genes in the population for a specific allele is frequency.

37. **(2)** According to the Hardy-Weinberg formula:

p = the frequency of the dominant gene (T)
q = the frequency of the recessive gene (t)

$p + q$ = 100% or 1.   and

p^2 = the homozygous dominant (TT)
$2\,pq$ = the hybrid dominant (Tt)
q^2 = the homozygous recessive (tt)

If q^2 = 25% or 0.25, then q = 0.50 then the frequency of the homozygous tasters (TT) = p^2
$p^2 = (.50)^2$ = 0.25 or 25%

INCORRECT CHOICES: (1), (3), and (4) are incorrect answers according to the calculations.

38. **(2)** Mutation and sexual reproduction increase variety in a population. INCORRECT CHOICES: (1), (3), and (4) Vegetative propagation, binary fission, and parthenogenesis are forms of asexual reproduction which produce offspring with traits identical to the parent.

39. **(2)** In the Hardy-Weinberg equation, $p^2 + 2pq + q^2 = 1$, $2pq$ represents the frequency for the heterozygous genotype and $2pq = 2 \times 0.5 \times 0.5 = 0.5 = 50\%$. INCORRECT CHOICES: (1) In this example, the frequency of the homozygous dominant genotype, p^2, and the frequency of the homozygous recessive genotype, q^2, both equal 25%. (3) The frequency of individuals with the dominant brown phenotype will equal 75%. (4) $p^2 + 2pq + q^2 = 100\%$.

40. **(3)** If brown mice are selected for mating, then the frequency of the dominant allele, p, will increase and q will decrease; as p increases, so will p^2. INCORRECT CHOICES: (1) If p increases, then $2pq$ cannot remain unchanged. (2) If brown mice are selected for mating, then the brown mice will pass more of their genes to the next generation than will the white mice, and the values of p and q will change. (4) q^2 will decrease rather than increase because there will be less chance of individuals receiving genes for white from both parents.

41. **(3)** Since the same kind of fossils appeared in old and new rocks, he inferred that the species had changed little over the years. INCORRECT CHOICES: (1) If the simplest fossil organisms appeared only in the oldest rocks, he might infer that the simplest organisms had evolved into more complex ones over the years. (2) If the simplest fossil organisms appeared only in the newest rocks, he might infer that the simplest organisms had evolved only recently. (4) If no fossil organisms of any kind appeared in the newest rocks, he might infer that the species had become extinct.

42. **(2)** Allele B went from a frequency of 0.99 in 1900 to a frequency of 0.10 in 1970 while allele b went from a frequency of 0.01 to a frequency of 0.90. INCORRECT CHOICES: (1) This represents the opposite of what occurred. (3) B is decreasing in frequency but b is increasing. (4) b is increasing in frequency but B is decreasing.

43. **(4)** A change in the environment favored the trait controlled by allele b leading to a selection of individuals with that trait and the consequent increase in the frequency of allele b. INCORRECT CHOICES: (1) Allele b may have existed in the gene pool as a result of mutation for an indefinite period of time before the change in the environment which favored the trait appeared. (2) If organisms carrying allele b had a reduced rate of

survival, then the frequency of allele *b* would decrease. (3) If environmental change favored allele *B*, then allele *B* would not have decreased in frequency.

44. **(1)** The frequency of homozygous (*BB*) individuals equals p^2; $p^2 = (0.10)(0.10) = 0.01 = 1\%$. INCORRECT CHOICES: (2) The frequency of heterozygous (*Bb*) individuals equals $2pq$; $2pq = 2(0.10)(0.90) = 0.18 = 18\%$. (3) The frequency of homozygous (*bb*) individuals equals q^2; $q^2 = (0.90)(0.90) = 0.81 = 81\%$. (4) All individuals in the population total 100%.

45. **(1)** A kingdom is the classification group including the largest number of different organisms. INCORRECT CHOICES: (2) Each kingdom is subdivided into phyla (singular: phylum). (3) Each phylum includes many genera (singular: genus). (4) A genus is subdivided into species.

46. **(2)** Paramecium and other protozoa are classified as part of the Protist kingdom. INCORRECT CHOICES: (1) The sponge is classified as an animal. (3) The Hydra is classified as an animal. (4) Moss is classified as a plant.

47. **(4)** Ideal classification is based on structural similarities among the organisms, and evolutionary relationships are based on structural similarities. INCORRECT CHOICES: (1) Very different organisms, the vulture and the tiger, have nutritional similarities but would not be classified in the same group. (2) Plants and animals both require water but would not be classified as similar organisms. (3) Elephants and large trees would not be placed in the same group because of size similarities.

48. **(4)** The term adaptive radiation means that the species will develop new species having a common ancestor. In Darwin's example the diversity of the environment allowed the finches to fill niches void of competition, and isolated from other members of the species. INCORRECT CHOICES: (1) The acquired characteristics theory of Lamarck was disproven by Weismann. (2) Blending inheritance or incomplete dominance was an idea of Mendel's. (3) Common ancestry is an idea that is given factual proof by evidence of evolution developed in such fields as biochemistry, genetics, and embryology.

49. **(1)** The kingdom represents the largest group of similar organisms. The system being referred to includes the following kingdoms: Plantae, Animala, and the final one is the Protista. INCORRECT CHOICES: (2), (3), and (4) These choices represent the smaller divisions of kingdom.

50. **(3)** The Protists are a group of plants and animals not easily classified into the other kingdoms. These usually include the unicellular species.

INCORRECT CHOICES: (1) Mollusks represent multicellular species characterized by soft, unsegmented bodies, and having three cell layers. (2) The fungi are plantlike and are included in the Protists. (4) The tracheophytes are the true land plants.

51. **(4)** Since autotrophs are self-feeders and carry on photosynthesis, it would be realistic to assume that as an end product of that process there would appear molecular oxygen. INCORRECT CHOICES: (1) To entertain the sequence of appearance suggested in this answer would be to show a lack of understanding of the organization of life namely from the simple to the complex. (2) Not all living things require oxygen to survive. The anaerobes do not require it and could have survived primitive earth conditions in the absence of oxygen. (3) Again, this choice is stating the reverse of the organization of life.

52. **(2)** According to the heterotroph hypothesis, the most primitive organism was a heterotroph which relied on anaerobic respiration as a source of energy in the absence of atmospheric oxygen. INCORRECT CHOICES: (1) Because the primitive atmosphere contained no oxygen, aerobic respiration was not possible. (3) and (4) The earliest organisms probably incorporated molecules from the seas as "food" and were therefore, heterotrophic, not autotrophic.

53. **(4)** Because asexual reproduction does not involve the fusion of two nuclei, the least changes would occur in the offspring. INCORRECT CHOICES: (1) Geographic isolation or confinement would result in changes to a species best suited to the isolated environment. (2) If the environments change, differing species will survive and reproduce through a selective and survival process. (3) Genetic recombination of inherited linkage groups occurring during crossing-over would result in more species changes than asexual reproduction.

54. **(2)** Population genetics is the study of the gene behavior or the application of Mendel's laws in populations or large groups of sexually reproducing organisms. INCORRECT CHOICES: (1) Adaptive radiation is the production of a number of different species from a single ancestral one. (3) The fraction of all members of the population which have a particular gene, expressed as a decimal is the gene frequency. (4) Variations are the differences which exist in the offspring of a particular species.

55. **(1)** According to the Hardy-Weinberg principle, not only will the pure dominant or homozygous individuals (p^2 or 0.25) display that trait, but the hybrid or heterozygous individuals ($2pq$, or $2 \times 0.5 \times 0.5$, or 0.5) display it also for a total of 75% of the population. INCORRECT CHOICES:

(2) Only q^2, or 0.25, or 25% of the population will display the recessive characteristic. (3) Seventy-five percent will display the dominant and 25% the recessive characteristics. (4) There will be 0.5 or $2pq$ hybrids and only 0.25 or q^2 recessive individuals for that trait.

56. **(3)** Since the publication of Darwin's work, scientists have discovered the genetic basis of evolutionary change. INCORRECT CHOICES: (1) Survival of the fittest is an important part of the theory of natural selection. (2) Usually more offspring are produced than can survive; therefore, there is competition within species. (4) Natural selection does not include Lamarck's theory of use and disuse.

57. **(1)** The term species means that the organisms would be able to mate and produce fertile offspring. If the organisms cannot do so, then they are not of the same species. INCORRECT CHOICES: (2) Even in two geographical areas, the organisms could be of the same species if they could produce fertile offspring. (3) The rate of mutations is not a criterion for being of the same species. (4) The genes do not necessarily have to be drawn from the same gene pool to be of the same species.

58. **(2)** Reproductive isolation of the species would indicate that two identical species separated by geography would reproduce and adapt to their specific environments and possibly develop new species. An example of this would be the squirrel population in the Grand Canyon. INCORRECT CHOICES: (1) Isolation of any kind would affect the species. (3) Mutations would have a higher occurrence in isolated species in that there would be no migration into or out of the population. (4) There could be no mixing of gene pools if the species were in isolation.

59. **(4)** Common ancestry is a theory of the interrelationships of different animals which is supported by the comparison of structures, embryo formation, and biochemistry. INCORRECT CHOICES: (1) Genetic dominance is concerned with phenotypes, or genotypes in the offspring. (2) Fossil formation, the study of the remains, imprints, and so forth, enables scientists to determine how long ago animals and plants lived on earth. (3) Geographic isolation in the evolutionary process is the isolation of a species with the eventual change that the species would undergo.

60. **(1)** Some mosquitoes had genetic traits which made them resistant to DDT. INCORRECT CHOICES: (2) DDT does not increase the chance of fertilization. (3) DDT is not a hormone. (4) DDT is an insecticide and not a source of food.

61. **(2)** According to Hardy-Weinberg, the gene pool remains stable if there is a large population, random mating, no migrations, and no mutations. INCORRECT CHOICES: (1), (3), and (4) Migrations, nonrandom mating, and a small population are factors which tend to change a gene pool.

62. **(4)** Populations exhibiting the greatest amount of variation are likely to have the traits needed to survive in a changing environment. INCORRECT CHOICES: (1), (2), and (3) These populations show less variation and therefore are less likely to survive drastic environmental change.

63. **(1)** The evolution of a species is most dramatically affected by changes in the environment. INCORRECT CHOICES: (2) Use and disuse affects the body cells not the sex cells. (3) Asexual reproduction produces organisms genetically identical and provides little variety for evolutionary change. (4) Vegetative propagation is a form of asexual reproduction.

64. **(4)** Darwin's theory offered a mechanism for the origin of the diverse organisms he observed. INCORRECT CHOICES: (1) Evolution is a theory not a fact and has not been proven. (2) Darwin's work preceded the discovery of genes or mutations. (3) Organisms of all sizes compete successfully in nature.

65. **(1)** Because of the absence of competition from antibiotic sensitive strains, the antibiotic resistant strains have become more prevalent. This is an example of natural selection. INCORRECT CHOICES: (2) Newer antibiotics should be more effective because sensitive organisms have not been eliminated. (3) Laboratory strains have an increased resistance due to natural selection. (4) If patients were less susceptible, there would be fewer outbreaks.

Questions 66–69 will be treated as one unit to avoid repetition of incorrect choices.

66. **(2)** A gene pool is the sum total of all the genes collectively present within a given population. Genes are composed of DNA and a protein.

67. **(1)** Adaptive radiation describes the production of a number of different species from a single ancestral one; each one of the new species fills an available niche in the environment.

68. **(3)** Those mosquitoes that were resistant to DNA were selected for in the environment. They reproduced and passed on the trait of resistance. Gradually the number of resistant mosquitoes increased. This is an example of natural selection.

69. **(5)** Similarities between embryos of groups of organisms strongly suggest a common ancestry. Both vertebrate embryos and mollusk and annelid larvae show these similarities. INCORRECT CHOICE: (4) Inability of organisms living at the same time in the same area to reproduce is referred to as reproductive isolation.

70. **(3)** $2pq$ refers to individuals with a gene for both the dominant trait (p) and the recessive trait (q). These individuals are heterozygous. INCORRECT CHOICES: (1) p^2 = homozygous dominant individuals. (2) q^2 = homozygous recessive individuals. (3) p = the dominant allele and q = recessive allele.

71. **(3)** Binary fission is a form of asexual reproduction which reproduces organisms identical to each other. This process would not produce the variations which are a basis for evolution. INCORRECT CHOICES: (1) Gene mutations are changes in the base code which can be inherited and produce variation. (2) The recombination of linkage groups during crossing-over and the combination of certain traits during fertilization both increase variation. (4) Crossing-over is the breakage of linkage groups and the exchange of these linkage groups between homologous chromosomes during meiosis. This also increases variation.

72. **(4)** q^2 = homozygous recessive allele = bb = .16
    then q = .4
    $p + q$ = 1.0
    then p = 1.0 − .4 = .6
    p^2 = homozygous dominant = BB
    p^2 = $(.6)^2$ = .36 = 35%

INCORRECT CHOICES: (1) and (2) Neither 4% nor 6% represent a correct answer. However 0.40 = q and 0.60 = p. (3) Sixteen percent represents the percentage of homozygous recessive individuals.

7

Plants and Animals in Their Environment (Ecology)

A. Ecology
   1. Physical factors in the environment
   2. Biotic factors in the environment

B. Biotic organization
   1. Population
   2. Community
   3. Ecosystem
      a. Interactions
         (1) Specific relationships
         (2) General relationships
            (a) Energy flow
            (b) Material cycles
      b. Maintenance
      c. Changes
         (1) Pioneer organisms
         (2) Climax community
   4. World biomes
      a. Terrestrial
      b. Aquatic
         (1) Marine
         (2) Fresh water
   5. Biosphere

C. Biosphere and Man
   1. Past and present
      a. Negative aspects
      b. Positive aspects
   2. The future

The following questions on Ecology have appeared on previous Regents Examinations.

Directions: Each question is followed by four choices.
Underline the correct choice.

1. During the 20th century, the greatest factor in upsetting the balance of nature has been the increase in the
   (1) salt content of the oceans
   (2) number of virus varieties on Earth
   (3) ability of humans to modify the environment
   (4) amount of solar radiation

Directions: Base your answers to questions 2 and 3 on the diagram below which shows a portion of a typical deciduous forest ecosystem which has been divided into zones in which different organisms predominate.

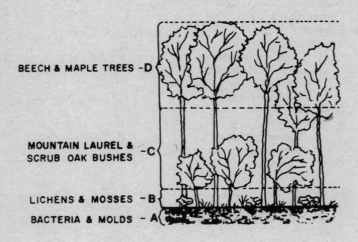

BEECH & MAPLE TREES -D

MOUNTAIN LAUREL & SCRUB OAK BUSHES -C

LICHENS & MOSSES -B

BACTERIA & MOLDS - A

2. The green plants adapted to surviving under the *lowest* light intensity would most likely be indicated by
   (1) *A*   (3) *C*
   (2) *B*   (4) *D*

3. Decomposers would be indicated by
   (1) A   (3) C
   (2) B   (4) D

4. Although the surface region of the oceans is rich in life, at greater depths
   the number of organisms decreases. This is primarily due to
   (1) decreasing light   (3) increasing salinity
   (2) decreasing density   (4) increasing temperature

5. In the food chain shown below, which organism represents a primary
   consumer?

   (1) grasshopper   (3) frog
   (2) grass   (4) snake

6. Which is a positive aspect of the human influence on the environment?
   (1) increased usage of DDT in agriculture
   (2) soil reclamation by reforestation
   (3) use of the automobile as primary transporation
   (4) urbanization of large amounts of farmland

7. In human females, the main function of the follicle-stimulating hormone
   (FSH) secreted by the pituitary gland is to
   (1) stimulate the adrenal glands to produce cortisone
   (2) stimulate activity in the ovaries
   (3) control the metabolism of calcium
   (4) regulate the rate of oxidation in the body

   Directions: Base your answers to questions 8 and 9 on
   the diagrams and the information on the following page.

A
BEAN

B
CHIMPANZEE

C
CHICKEN

D
AMEBA

8. Which organisms were produced as a result of fertilization?
   (1) A, B, and C, only   (3) C and D, only
   (2) B and C, only   (4) B, C, and D, only

9. Structures which function in the storage of food to be used by growing embryonic cells are indicated by
   (1) 1 and 3   (3) 2 and 4
   (2) 2 and 3   (4) 3 and 4

10. A package of seeds is opened and the seeds are planted. What is the most direct sequence of events that follow?
    (1) germination → differentiation → pollination → fruit formation
    (2) fruit formation → pollination → differentiation → germination
    (3) differentiation → pollination → germination → fruit formation
    (4) pollination → fruit formation → germination → differentiation

   Directions: Base your answers to questions 11 through 13 on your knowledge of biology and the information below.

   A biologist cut a flap of ectoderm from the top of a developing embryo. He did not remove the piece of ectoderm but just folded it back. Then he cut out the mesoderm underneath and completely removed it. He folded the flap of ectoderm back in place. The ectoderm healed, however, a complete nervous system did not develop.

11. This experiment was most likely performed immediately after
    (1) cleavage   (3) fertilization
    (2) gestation   (4) gastrulation

12. This experiment interfered with the process of
    (1) differentiation
    (2) zygote formation
    (3) cleavage
    (4) ovulation

13. This experiment demonstrates that the
    (1) ectoderm is solely responsible for the development of the nervous system
    (2) nervous system is destroyed during surgical operations
    (3) mesoderm influences the development of the nervous system
    (4) digestive enzymes have a major role in the development of embryonic layers

14. In a developing embryo, the mesoderm layer would normally give rise to
    (1) epidermal tissue
    (2) skeletal tissue
    (3) digestive tract lining
    (4) respiratory tract lining

15. In plants, two areas of rapid mitotic division are the
    (1) epidermis and xylem
    (2) xylem and phloem
    (3) root tip and cambium
    (4) cambium and phloem

16. If the first stage of an uninterrupted human menstrual cycle is the follicle stage, then the last stage would include the
    (1) formation of sperm cells in the testis
    (2) release of a mature egg
    (3) buildup of the uterine lining
    (4) shedding of the uterine lining

17. In a particular area, the living organisms and nonliving environment function together as
    (1) a population
    (2) a community
    (3) an ecosystem
    (4) a species

18. If line A in the diagram below represents a population of hawks in a community, then what would most likely be represented by line B?

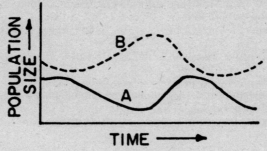

(1) the dominant trees in that community
(2) a population with which the hawks have a mutualistic relationship
(3) variations in the numbers of producers in that community
(4) a population on which the hawks prey

19. Compared to large land areas, large bodies of water such as oceans absorb heat
    (1) much faster and release it much faster
    (2) faster and release it more slowly
    (3) more slowly and release it at a slower rate
    (4) more slowly and release it faster

20. What is the usual order of the biomes from the Equator to the North Pole at sea level?
    (1) taiga, tropical forest, temperate deciduous forest, tundra
    (2) temperate deciduous forest, tropical forest, taiga, tundra
    (3) tropical forest, temperate deciduous forest, taiga, tundra
    (4) tropical forest, taiga, temperate deciduous forest, tundra

21. Which type of organism is *not* shown in the following representation of a food chain?

    grass → mouse → snake → hawk

    (1) herbivore   (3) producer
    (2) decomposer   (4) carnivore

22. A lake contains minnows, mosquito larvae, sunfish, algae, and pike. Which of these organisms would probably be present in the largest numbers?
    (1) minnows   (3) sunfish
    (2) larvae   (4) algae

23. Which climax vegetation indicates a taiga?
    (1) coniferous trees   (3) annual grasses
    (2) cactus and mesquite   (4) lichens and mosses

24. Most of New York State is located in which land biome?
    (1) tropical rain forest   (3) grassland
    (2) tundra   (4) temperate deciduous forest

25. Which of the following is the most ecologically promising method of insect control?
    (1) interference with insect reproductive processes
    (2) stronger insecticides designed to kill higher percentages of insects
    (3) physical barriers to insect pests
    (4) draining marshes and other insect habitats

26. Which is an example of a biological control of a pest species?
    (1) DDT was used to destroy the red mite.
    (2) Most of the predators of a deer population were destroyed by humans.
    (3) Gypsy moth larvae (a tree defoliator) are destroyed by beetle predators which were cultured and released.
    (4) Drugs are used in the control of certain pathogenic bacteria.

Directions (27–31): For each description in questions through 31, select the biome, *chosen from the list below,* that is most closely associated with that description.

Biomes

(1) Desert
(2) Grassland
(3) Taiga
(4) Temperate deciduous forest
(5) Tundra

27. This area has a short growing season and low precipitation, mostly in the form of snow. The soil is frozen permanently and vegetation includes lichens and mosses.

28. This area has 10 to 30 inches of rainfall annually. The growing season does not produce trees, but the soil is rich and well-suited for growing domesticated plants such as wheat and corn. Grazing animals are found here.

29. There are many lakes in this area and vegetation is coniferous forests composed mainly of spruce and fir. There are many large animals like bear and deer.

30. This area has broad-leaved trees which shed their leaves in the fall. Winters are fairly cold, and the summers are warm with well-distributed rainfall.

31. This area has a low annual rainfall and a rapid rate of evaporation. In order for plants to survive, they must be adapted to conserve moisture. Animals here are active mainly at night.

32. The whole area of the Earth where ecosystems operate is known as
    (1) a biome   (3) the biosphere
    (2) a community   (4) the atmosphere

33. Which biome is characterized by its ability to absorb and hold large quantities of solar heat that helps to regulate the Earth's temperature?
    (1) desert biome
    (2) marine biome
    (3) grassland biome
    (4) taiga biome

34. The biotic and abiotic factors interacting with each other in a pond form
    (1) an ecosystem
    (2) a community
    (3) a population
    (4) a food web

35. Horses are herbivores and, as a result, can be classified as
    (1) omnivores
    (2) primary consumers
    (3) secondary consumers
    (4) decomposers

36. If a certain type of poison were to destroy nitrogen-fixing bacteria, the most immediate result would be
    (1) a decrease in the percentage of atmospheric nitrogen
    (2) a decrease in the nitrate concentration in legumes
    (3) an increase in the percentage of atmospheric CO_2
    (4) an increased number of healthier legumes

37. The graph below shows the effect of a factor on the photosynthetic rate of the green marine algae *Enteromorpha linza*. Which is most likely represented by factor X?

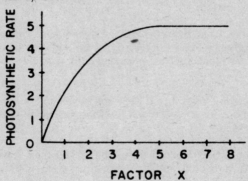

    (1) light intensity
    (2) water concentration
    (3) competition level
    (4) substratum type

38. A coniferous forest would be *least* likely to appear within the
    (1) United States
    (2) Arctic Circle
    (3) Canadian Provinces
    (4) U.S.S.R.

39. The trees in a forest aid in controlling floods chiefly because their
    (1) branches store water in the form of sap
    (2) leaves absorb moisture from the air
    (3) root systems retain the soil substratum
    (4) stems serve as reservoirs for food

40. Aerobic organisms are dependent on autotrophs. One reason for this
    dependency is that most autotrophs provide the aerobic organisms with
    (1) oxygen   (3) nitrogen gas
    (2) carbon dioxide   (4) hydrogen

    Directions (41–45): For each statement in questions 41
    through 45, select the biome, *chosen from the list below,*
    that is best described by that statement.

    Biomes

        (1) Tundra
        (2) Temperate deciduous forest
        (3) Tropical rain forest
        (4) Desert
        (4) Marine
        (6) Taiga

41. Cold winters and warm summers characterize the biome where most
    trees lose their leaves in the winter.

42. Plants have extensive root systems for conserving water.

43. Extreme amounts of rainfall and very hot temperatures characterize this
    biome of dense vegetation.

44. Very cold temperatures keep the ground permanently frozen (permafrost)
    a few feet below the surface even in the summer, and mosses and lichens
    are common.

45. This biome has the most stable habitat for most organisms and contains
    a relatively constant supply of nutrients.

46. Which organism is a near-extinct species?
    (1) Japanese beetle   (3) blue whale
    (2) dodo bird   (4) passenger pigeon

47. What has been the main cause of the upset of the delicate balance of nature in the biosphere?
    (1) use of biodegradable detergents
    (2) uncontrolled use of guns
    (3) large numbers of people
    (4) large numbers of producers

48. In order to avoid predators, the clown fish hides unharmed in the stinging tentacles of the sea anemone. The clown fish attracts food to the sea anemone. This is an example of a type of relationship known as
    (1) mutualism   (3) predator-prey
    (2) commensalism   (4) parasitism

49. A slab of bare rock is covered with lichens. In time, mosses cover the rock, followed by grasses, and finally by small shrubs and tree saplings. In this example, the lichens represent
    (1) a climax community   (3) secondary consumers
    (2) a dominant species   (4) pioneer organisms

Directions: Base your answers to questions 50 through 52 on the graph below and on your knowledge of biology. The graph represents the relationship between the capacity of the range (number of deer that could be supported by the range), the number of deer actually living on the range, and time.

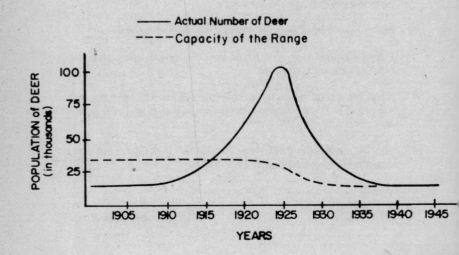

50. In what year was the number of deer living on the range equal to the capacity of the range?
    (1) 1905   (3) 1920
    (2) 1915   (4) 1930

51. What is the most likely reason why the capacity of the range to support deer decreased between 1920 and 1930?
    (1) The deer population became too large.
    (2) The number of predators increased between 1915 and 1925.
    (3) The deer population decreased in 1919.
    (4) An unusually cold winter occurred in 1918.

52. What might be one reason why the number of deer began to increase in 1910?
    (1) The deer's natural enemies were killed.
    (2) The capacity of the range increased.
    (3) The available vegetation of the area decreased.
    (4) The winter was longer than normal in 1905.

53. The portion of the Earth in which all ecosystems operate is known as the
    (1) tropics   (3) temperate zone
    (2) community   (4) biosphere

54. Which is true of the major land biomes?
    (1) They are characterized by the animals living in the region.
    (2) They are unaffected by major climatic changes.
    (3) They are named by the climax vegetation in the region.
    (4) They are located predominantly at lower latitudes and altitudes.

Directions: Base your answers to questions 55 and 56 on the diagram below which represents a cross section of a marine biome and on your knowledge of biology.

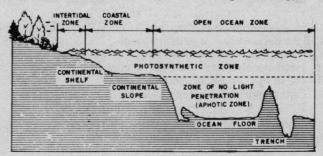

55. Which zone usually has the highest food production per unit volume?
    (1) ocean floor zone
    (3) open ocean zone
    (2) aphotic zone
    (4) coastal zone

56. Compared to similar land environments farther inland, the landmass near this marine biome would tend to be
    (1) warmer in winter, due to heat retention by the water
    (2) less productive, due to high humidity near the shore
    (3) colder in winter, due to presence of salt in the water
    (4) richer in plant species, due to higher fertility of the soil

57. If a person traveled south from the Arctic Circle to the Equator, what would be the most probable sequence of land biomes he would pass through?
    (1) temperate forest → taiga → tundra → tropical forest
    (2) taiga → tundra → temperate forest → tropical forest
    (3) tundra → tropical forest → taiga → temperate forest
    (4) tundra → taiga → temperate forest → tropical forest

58. What is the function of the placenta in a mammal?
    (1) It surrounds the embryo and protects it from shock.
    (2) It allows the mixing of the maternal and fetal blood.
    (3) It permits the passage of nutrients and oxygen from the mother to the fetus.
    (4) It replaces the heart of the fetus until the fetus is born.

59. A mold which grows on the body of a dead earthworm is a
    (1) parasite
    (3) herbivore
    (2) saprophyte
    (4) predator

60. Over very long periods of time, many areas of barren rock may become covered with dense forests. This phenomenon is known as
    (1) biological control
    (3) population fluctuation
    (2) species migration
    (4) ecological succession

61. Which statement best explains why an ecosystem could no longer exist if all the decomposers were eliminated?
    (1) The producers would outnumber the consumers.
    (2) The materials could no longer be recycled.
    (3) The oxygen supply would be depleted.
    (4) Carbon dioxide would not be used at a fast enough rate.

62. In the biosphere, what are some major abiotic factors which determine the distribution and types of plant communities?

(1) temperature, sunlight, and rainfall
(2) humidity, location, and humans
(3) soil type, soil bacteria, and soil water
(4) insects, carbon dioxide, and nitrogen in the air

63. All the plants, animals, and protists interacting in a given environment make up a
(1) population
(2) species
(3) community
(4) niche

64. In which example of a nutritional relationship is an organism harmed?
(1) alga and fungus in a lichen
(2) nitrogen-fixing bacteria and clover
(3) remora and a shark
(4) athlete's foot fungus and humans

65. Many different species of organisms interacting in a particular environment are an example of a
(1) population
(2) biosphere
(3) community
(4) biome

66. An organism is introduced into a community and occupies the same ecological niche as another organism. Both organisms might be expected to
(1) reverse their material cycles
(2) compete for the niche
(3) develop climax populations
(4) become extinct after a few generations

67. Which condition would most likely upset the dynamic equilibrium of an ecosystem?
(1) a constant source of energy entering the environment
(2) a cycling of elements between organisms and the environment
(3) organisms capable of incorporating radiant energy into organic compounds
(4) a greater number of heterotrophs than autotrophs

68. In a study made over a period of years in a certain part of the country, the research showed that there was a low amount of rainfall, a wide seasonal variation in temperature, and short periods of daylight. These environmental factors are
(1) abiotic factors of little importance to biotic factors
(2) abiotic factors that limit the type of organisms present in the area
(3) biotic factors important to saprophytes in the area
(4) biotic factors which are affected by the abiotic factors

69. Which list of ecological terms correctly illustrates the levels of biotic organization from the simplest to the most complex?
    (1) organism — community — ecosystem — population — biosphere
    (2) population — community — organism — biosphere — ecosystem
    (3) ecosystem — organism — population — community — biosphere
    (4) organism — population — community — ecosystem — biosphere

70. The presence of nitrogen-fixing bacteria in nodules on the roots of legumes such as the peanut plant illustrates an association known as
    (1) commensalism   (3) parasitism
    (2) mutualism   (4) saprophytism

71. Which pioneer organisms are most likely to occupy the bare rock habitat of a mountaintop?
    (1) lichens   (3) legumes
    (2) grasses   (4) hemlock trees

72. Which is a biotic factor in an environment?
    (1) the average number of hours of sunlight reaching a field
    (2) the variations in daily temperature surrounding a cactus
    (3) the number of trees competing for water and light
    (4) the quantity of oxygen in a pond

73. Organic matter from dead plants and animals is broken down and returned to the soil by the action of
    (1) decomposers   (3) producers
    (2) herbivores   (4) parasites

74. The chief competitors for the food supply used by humans are
    (1) rabbits   (3) earthworms
    (2) insects   (4) rats

75. A hawk is an example of a
    (1) parasite   (3) carnivore
    (2) herbivore   (4) saprophyte

76. In a community, the greatest amount of available stored food and energy is found in the
    (1) producers   (3) secondary consumers
    (2) primary consumers   (4) decomposers

77. Why is the use of pesticides such as DDT being discouraged in many countries?

(1) Crop production may increase because of their use.
(2) They disrupt food webs.
(3) All insects are immune to them.
(4) They do not enter material cycles.

78. In which biome are the abiotic factors most constant?
 (1) ocean
 (2) desert
 (3) deciduous forest
 (4) tundra

79. Forest preservation helps to prevent flooding because the
 (1) leaves of trees carry on photosynthesis
 (2) leaves of trees carry on transpiration
 (3) roots of trees hold the soil particles together
 (4) xylem ducts conduct water to the leaves

80. In which areas does the greatest amount of photosynthesis occur?
 (1) mountains
 (2) oceans
 (3) deserts
 (4) tundras

Directions: Base your answers to questions 81 through 83 on the information below and on your knowledge of biology.

An aquarium container is filled with water and colonies of aquatic and animals. Various protists are added and the aquarium is then sealed and placed on a window ledge. After a period of time the aquarium appears to reach a state of balance.

81. The oxygen content of the tank is maintained by the
 (1) autotrophs
 (2) heterotrophs
 (3) fungi
 (4) carnivores

82. The energy needed to maintain this ecosystem originates from the
 (1) fish
 (2) green plants
 (3) water
 (4) Sun

83. Which group of organisms in the aquarium contains the largest amount of energy?
 (1) primary consumers
 (2) secondary consumers
 (3) producers
 (4) herbivores

84. Two organisms are placed in the same species if they
    (1) are able to mate and produce fertile offspring
    (2) are able to share similar environments
    (3) both require the same food materials
    (4) both have structures for aerobic respiration

85. Before it was banned, the insecticide DDT was used to combat an organism called the red mite. An unexpected result of the use of DDT was that the population of the red mite increased rather than decreased, while the population of insect predators of the red mite declined. What is the most probable explanation of this phenomenon?
    (1) Part of the red mite population was resistant to DDT and its predators were not.
    (2) DDT is highly general in the kinds of insects it affects, killing both beneficial and harmful species.
    (3) The red mite population could use DDT as a nutrient, while the predators could not.
    (4) DDT triggered a mutation in the red mite population, making it immune to the effects of the chemical.

86. Some bacteria are classified as saprophytes because they are organisms which
    (1) feed on other living things
    (2) feed on dead organic matter
    (3) manufacture food by photosynthesis
    (4) contain vascular bundles

87. Generally, an increase in altitude has the same effect on the habitat of organisms as
    (1) an increase in latitude   (3) a decrease in available light
    (2) an increase in moisture   (4) a decrease in longitude

88. Of the following, the greatest amount of the Earth's food production is thought to occur in
    (1) coastal ocean waters   (3) taiga forests
    (2) desert biomes   (4) tundra biomes

89. Which is the most stable biome?
    (1) desert   (3) taiga
    (2) temperate deciduous forest (4) marine

90. Which are usually the last types of plants to appear in a forest succession?
    (1) grasses   (3) pioneer plants
    (2) mosses   (4) climax plants

Answers

1. 3	19. 3	37. 1	55. 4	73. 1
2. 2	20. 3	38. 2	56. 1	74. 2
3. 1	21. 2	39. 3	57. 4	75. 3
4. 1	22. 4	40. 1	58. 3	76. 1
5. 1	23. 1	41. 2	59. 2	77. 2
6. 2	24. 4	42. 4	60. 4	78. 1
7. 2	25. 1.	43. 3	61. 2	79. 3
8. 1	26. 3	44. 1	62. 1	80. 2
9. 3	27. 5	45. 5	63. 3	81. 1
10. 1	28. 2	46. 3	64. 4	82. 4
11. 4	29. 3	47. 3	65. 3	83. 3
12. 1	30. 4	48. 1	66. 2	84. 1
13. 3	31. 1	49. 4	67. 4	85. 1
14. 2	32. 3	50. 2	68. 2	86. 2
15. 3	33. 2	51. 1	69. 4	87. 1
16. 4	34. 1	52. 1	70. 2	88. 1
17. 3	35. 2	53. 4	71. 1	89. 4
18. 4	36. 2	54. 3	72. 3	90. 4

Explanatory Answers

1. **(3)** Human beings' ability to modify the environment has been the single greatest factor which has had a negative effect on the balance of nature. INCORRECT CHOICES: (1) The salt content of the oceans is very stable. (2) The number of virus varieties is not a significant factor in upsetting the balance of nature. (4) The amount of solar radiation is also relatively constant.

2. **(2)** The lichens and mosses, living under the canopy of leaves on the trees and shrubs have adapted to low intensity light. INCORRECT CHOICES:

(1) Bacteria and molds are nonphotosynthetic and therefore low light intensity would not affect them. (3) The mountain laurel and oak bushes survive under low light intensity but not the lowest. (4) The deciduous trees receive the highest light intensity.

3. **(1)** The bacteria and molds are decomposers in nature. INCORRECT CHOICES: (2), (3), and (4) All of these choices are photosynthetic.

4. **(1)** The amount of light available is the most significant factor in determining the amount of photosynthesis and, therefore, the number of organisms living at a particular depth. INCORRECT CHOICES: (2) and (3) Density and salinity do change with depth, but these are not the most important factor in providing food for organisms. (4) Temperature decreases at greater depths, but this is not as important as the amount of available light.

5. **(1)** A primary consumer feeds on the producer organisms. In this diagram the grasshopper is the only primary consumer shown. INCORRECT CHOICES: (2) The grass is the producer oganism. (3) The frog is the secondary consumer. (4) The snake is not a primary consumer; it is a herbivore.

6. **(2)** Soil reclamation is needed as a human influence so that the human population can provide growing space for its own needs and provide space for the expansion of plant and animal species. Thus it represents a positive influence. INCORRECT CHOICES: (1), (3), and (4) These choices represent negative influences of human influence.

7. **(2)** Follicles are structures located in the ovaries in which egg cells mature. FSH stimulates these structures for the maturation of eggs. INCORRECT CHOICES: (1) The adrenal glands are stimulated by ACTH from the pituitary gland. (3) Parathormone is produced by the parathyroid glands to control or regulate the metabolism of calcium. (4) Thyroxin is released by the thyroid gland for the regulation of the rate of oxidation in the body.

8. **(1)** The bean, chimpanzee, and chicken embryos are the results of the union of two different sex cells. INCORRECT CHOICES: (2) The bean embryo is also a result of fertilization. (3) New amebae are the result of asexual reproduction—fission. (4) The ameba is not the result of fertilization.

9. **(3)** The cotyledon of the bean and the yolk sac of the chicken egg contain food for the developing embryonic cells. INCORRECT CHOICES: (1), (2), and (4) Structure number 1 will develop into bean plant leaves and structure number 3 is the uterus in which the embryo is attached.

10. **(1)** The embryo and young plant develop from the seed during germination. The cells of the embryo and young plant differentiate to form structures including flowers which produce pollen. The pollen goes from anther sac to stigma of the pistil (blossoms) after which fertizilation occurs resulting in the formation of fruit. INCORRECT CHOICES: (2), (3), (4) Germination, or the growth of the embryo or young plant, occurs first.

11. **(4)** The nervous system, which forms from the ectoderm, did not fully develop because of the partial inhibiting of the induction of the mesoderm cells to activate the entire ectoderm of the hollow three-layered gastrula. INCORRECT CHOICES: (1) Cleavage is the mitotic development of the zygote to form the blastula. (2) Gestation is the process of growth and development of an embryo from the zygote to hatching or birth. (3) Fertilization is the union of two heterozygous sex cells.

12. **(1)** This experiment interfered with differentiation whereby embryonic cells were developing into specific nerve tissue cells of a many celled organism. INCORRECT CHOICES: (2) A zygote was formed prior to the gastrula. (3) Cleavage occurred prior to and after the experiment. (4) Ovulation is the release of eggs from the ovary of a female.

13. **(3)** The mesoderm influences the development of tissues of the ectoderm by induction. INCORRECT CHOICES: (1) Each of the three layers of the gastrula have an effect on one another. (2) The entire nervous system is not destroyed during surgical operations. (4) Digestive enzymes have a major role in the hydrolysis of food.

14. **(2)** Skeletal tissue or bones are formed from the mesoderm layer. INCORRECT CHOICES: (1) Epidermal tissue is formed from the ectoderm layer. (3) and (4) Digestive tract lining and respiratory tract lining are formed from the endoderm layer.

15. **(3)** The root tip and cambium cells are undifferentiated meristem tissue specialized for growth. INCORRECT CHOICES: (1) Epidermal tissue covers and protects inner structures. Xylem tissue conducts water from the roots through the stems to the leaves. (2) and (4) The phloem tissue conducts the nutrients formed in the leaves to the stems and roots.

16. **(4)** The uterine lining which built up during the menstrual cycle is shed in the last phase of menstruation. INCORRECT CHOICES: (1) The formation of sperm occurs during spermatogenesis in males. (2) The mature egg is released during ovulation which occurs prior to the last stage of the menstrual cycle. (3) The building up of the uterine lining occurs during the early stages of the menstrual cycle.

17. **(3)** An ecosystem is composed of the functioning relationships of living organisms and the nonliving environment. INCORRECT CHOICES: (1) A population is all the individuals of a given species in a particular area. (2) A community is composed of all the living organisms in a given area. (4) A species is a portion of the living organisms in an area which can produce offspring.

18. **(4)** When the population of hawks increases, the population of its prey decreases and vice versa. INCORRECT CHOICES: (1) There is no apparent relationship between the dominant trees and the hawks. It would seem that the greater the number of trees, the greater the number of hawks should be present, since the trees provide places for them to roost. (2) The hawks seem to have a predator–prey relationship with the other organisms. (3) Usually, there are no one-step relationships between carnivores and producers (green plants).

19. **(3)** Water, by its chemical nature, requires large additions of heat to increase its molecular temperature. The molecular bonds formed as a result of this absorption are strong, and we can then realize why water retains its heat longer. INCORRECT CHOICES: (1) This choice is the direct opposite of the correct choice. (2) Water absorbs heat slower, not faster. (4) Water releases its heat slower, not faster.

20. **(3)** This choice presents the correct order. The tropical forest is at the equator (hotter) and the temperature cools down progressively between the equator and the North Pole ending in the tundra biome. INCORRECT CHOICES: (1) The taiga biome is not in order. (2) The tropical forest and taiga biomes are out of order. (4) Again the taiga biome is out of order.

21. **(2)** A decomposer would be a type or types of bacteria. INCORRECT CHOICES: (1) A herbivore is an organism which uses plants for its food. (3) A producer is an organism which makes its own food (an autotroph). (4) A carnivore uses fleshy food or meat for food.

22. **(4)** The algae will be found in largest numbers. They are producers (make their own food) and provide the base for the food chain and the pyramid of energy. INCORRECT CHOICES: (1) and (3) Minnows and starfish are consumers of the algae and would be found in less numbers. (2) The larva is the metamorphic stage between the egg and adult mosquito which lives in water. Larvae would be less in number than the algae.

23. **(1)** A taiga is a land surface with cold temperatures, a summer growing season, and where coniferous tress predominate. INCORRECT CHOICES: (2) Cactus and mesquite are found growing in a desert area. (3) Annual

grasses grow abundantly on the grassland area of North America, just east of the Rocky Mountains where there is little rain. (4) Lichens and mosses would be found in bare rock areas located in a temperate climate.

24. (4) Most of New York State is located in a land biome called a temperate deciduous forest with moderate rainfall, cold winters, and warm summers. INCORRECT CHOICES: (1) Tropical rain forests are areas of constant warmth and abundant water located in Central and South America, Southeast Asia, and West Africa. (2) The tundra is a land zone surrounding the Arctic ocean which is frozen to a depth of a few meters. (3) The major grasslands are located in a region just east of the Rocky Mountains in the Plains States.

25. (1) Interfering with the insect reproductive process by biological or selected chemical methods can be strictly controlled and will contribute least to the pollution of the environment. INCORRECT CHOICES: (2) The stronger insecticides many times have harmful effects on organisms other than the target organisms. (3) Physical barriers or quarantines have not proven as effective as the biological controls. Too many organisms escape the quarantine area. (4) Draining marshes and other insect habitats destroys those places for most organisms.

26. (3) Of the choices, this one is the only one in which we see the control of a pest by a natural predator. INCORRECT CHOICES: (1) DDT is a chemical control. (2) Humans in this case destroyed the natural predators of the deer, which is the opposite of the case in the question. (4) Drugs represent chemical control.

27. (5) The characteristics listed could only be for the tundra. INCORRECT CHOICES: (1) The desert is characterized by extremely low precipitation and very rapid evaporation. Plants survive by having adapted to water conservation and animals are active only at night. (2) The grasslands have a higher rate of precipitation, few trees, and rich soil well suited for grass crops. (3) The taiga has many lakes, cool temperatures, evergreen vegetation, and large animals. (4) The temperate deciduous forest is characterized by trees having broad leaves which are shed in the fall, cold winters, and warm summers.

28. (2) The description fits only the grasslands. INCORRECT CHOICES: (1), (3), (4), and (5) Refer to number 27.

29. (3) This description will fit only the taiga. INCORRECT CHOICES: (1), (2), (4), and (5) Refer to number 27.

30. **(4)** This description will fit only the temperate deciduous forest. INCORRECT CHOICES: (1), (2), (3), and (5) Refer to number 27.

31. **(1)** This description will fit only the desert biome. INCORRECT CHOICES: (2), (3), (4), and (5) Refer to number 27.

32. **(3)** The biosphere includes all the ecosystems. INCORRECT CHOICES: (1) A biome is a climax community of flora and fauna characteristic of one area of climate. (2) A community is a group of different species all interacting. (4) The atmosphere is the envelope of gases surrounding the earth.

33. **(2)** The water absorbs tremendous solar energy before increasing its own temperature, and retains the heat for a much longer period than other biomes. INCORRECT CHOICES: (1) The desert reflects much of the incoming radiation. (3) The grasslands absorb much of the energy and use it for photosynthesis. (4) The taiga has very little material to absorb and retain the radiant energy.

34. **(1)** The definition of an ecosystem is the interrelationship of the biotic and abiotic factors. INCORRECT CHOICES: (2) A community is a group of different species interacting. (3) A population is a group of the same species in an area. (4) A food web depicts the nutritional relationships in a community.

35. **(2)** Organisms that feed directly on green plants (herbivores are plant-eating animals) are known as primary consumers. INCORRECT CHOICES: (1) Omnivores are animals that eat both plants and animals. (3) Secondary consumers are animals that prey upon primary consumers. (4) Decomposers are organisms which accomplish the breakdown of wastes and dead organisms to simpler compounds.

36. **(2)** Nitrogen-fixing bacteria produce nitrates from gaseous nitrogen; since certain nitrogen-fixing bacteria are mutually symbiotic with legumes, the nitrate concentration in these legumes would decrease if the bacteria were destroyed. INCORRECT CHOICES: (1) The destruction of nitrogen-fixing bacteria would not decrease the percentage of atmospheric nitrogen because normal activity of these bacteria changes nitrogen to nitrates. (3) The destruction of nitrogen-fixing bacteria would not increase the percentage of atmospheric CO_2. (4) Legumes depend upon their mutual symbionts, the nitrogen-fixing bacteria, as a source of nitrates.

37. **(1)** Light intensity is a physical factor in the environment with an effect on photosynthetic rate like that shown; as the light intensity increases, the rate of photosynthesis increases up to a point after which the rate stays the same. INCORRECT CHOICES: (2) Providing there is sufficient water available to serve as a raw material for photosynthesis, the concentration of water will not have as great an effect on the photosynthetic rate as the amount of light. (3) An increase in competition for available resources would probably cause a decrease in photosynthetic rate. (4) Changes in substratum type are unlikely to affect rate of photosynthesis.

38. **(2)** The temperature is too low and the rainfall too small to support the growth of a coniferous forest within the Arctic Circle. INCORRECT CHOICES: (1), (3), and (4) The United States, the Canadian Provinces, and the USSR all have land masses within latitudes where a coniferous forest or taiga can exist.

39. **(3)** The roots form a web network which hold the soil particles in place. INCORRECT CHOICES: (1) Sugar is stored in the form of sap. (2) Leaves lose water by transpiration. (4) Stems usually do not keep soil in place and therefore do not control floods.

40. **(1)** One of the by-products of photosynthesis by autotrophs is oxygen which is used in aerobic respiration. INCORRECT CHOICES: (2) Carbon dioxide is formed from aerobic respiration. (3) and (4) Nitrogen gas and hydrogen gas are not necessary for aerobic respiration.

41. **(2)** Cold winters and warm summers characterize a temperate climate; in a deciduous forest most trees lose their leaves in the winter. INCORRECT CHOICES: (1) Very cold temperatures are characteristic of tundra. (3) Very hot temperatures characterize the tropical rain forest. (4) Very little rainfall is characteristic of desert. (5) Marine biomes are found in the oceans. (6) Taiga is characterized by coniferous, evergreen trees such as spruce, pine, and fir.

42. **(4)** In the desert, where rainfall is minimal, plants have extensive root systems for conserving water. INCORRECT CHOICES: (1) In the tundra, the ground is permanently frozen a few feet below the surface even in the summer. (2) Cold winters, warm summers, and moderate rainfall characterize a temperate deciduous forest. (3) Extreme amounts of rainfall are found in a tropical rain forest. (5) Marine biomes are found in the ocean. (6) The taiga is a coniferous forest.

43. **(3)** Extreme amounts of rainfall and very hot temperatures make possible the growth of dense vegetation in the tropical rain forest. INCORRECT

CHOICES: (1) Very cold temperatures are characteristic of tundra. (2) Cold winters and warm summers characterize the temperate deciduous forest. (4) Very little rainfall is characteristic of desert. (5) Marine biomes are found in the ocean. (6) The taiga is a coniferous forest.

44. **(1)** Mosses and lichens are dominant organisms in the cold tundra where the ground is permanently frozen a few feet below the surface even in summer. INCORRECT CHOICES: (2) Cold winters and warm summers characterize the temperate deciduous forest. (3) Very hot temperatures characterize the tropical rain forest. (4) Very little rainfall is characteristic of the desert. (5) Marine biomes are found in the ocean. (6) Coniferous trees such as pine, spruce, and fir are common in the taiga.

45. **(5)** The marine biome is the most stable habitat and contains a relatively constant supply of nutrients. INCORRECT CHOICES: (1) Very cold temperatures maintain permafrost in the tundra. (2) Seasonal variation in temperature is characteristic of temperate deciduous forest. (3) Extreme amounts of rainfall and high temperatures are characteristic of tropical rain forest. (4) Very little rainfall is typical of the desert biome. (6) The taiga is cold and supports coniferous forests of spruce, pine, and fir.

46. **(3)** There are very few blue whales left alive on our planet. They are hunted for food and their oils. INCORRECT CHOICES: (1) There are an abundant number of Japanese beetles on our planet and the population seems to be increasing. (2) and (4) The dodo bird and the passenger pigeon are no longer alive on our planet, therefore, they have become extinct.

47. **(3)** The large numbers of people have far outstripped the food-producing capacity of many ecosystems of the world thus upsetting the balance of nature. INCORRECT CHOICES: (1) Biodegradable detergents are those which can be broken down into reusable nutrients by decomposers. (2) The use of guns has led to the extinction or near extinction of many animal species, but this is only one effect of the rapidly increasing human population. (4) In most natural communities, producers will be found in the largest numbers.

48. **(1)** Both organisms benefit from the presence of the other. INCORRECT CHOICES: (2) Commensalism results when two organisms live in the same area and one organism benefits from the relationship and the other is not harmed by it. (3) A predator–prey relationship exists when one organism, the predator, uses the other organism, the prey, for its food. (4) A parasite is an organism which lives on or in a host from which it takes its food or nourishment.

49. **(4)** Pioneer organisms are the first plants to populate a given location. INCORRECT CHOICES: (1) A climax community is a stable one which can perpetuate itself until major climatic, geologic, or biotic changes alter or destroy it. (2) A dominant species is one which exerts control over the other species present. (3) Secondary consumers are animals which prey upon the primary consumers.

50. **(2)** On the graph, the actual number of deer and the capacity of the range intersect in the year 1915. INCORRECT CHOICES: (1) In 1905, the capacity of the range was greater. (3) and (4) In 1920 and 1930, the actual number of deer was greater.

51. **(1)** Overgrazing decreased the support capacity of the range. INCORRECT CHOICES: (2) If the number of predators increased, the number of deer would decrease. (3) The number of deer increased in 1919. (4) Weather is not indicated on the graph and the number of deer increased.

52. **(1)** A decrease in animals that prey on the deer would cause the deer population to increase. INCORRECT CHOICES: (2) The capacity of the range decreased. (3) A decrease in vegetation would cause a decrease in the deer population. (4) The length of winter weather is not indicated on the graph.

53. **(4)** The biosphere is the portion of the planet in which ecosystems operate. INCORRECT CHOICES: (1) The tropics are areas near the equator at sea level with high temperature and humidity. (2) A community consists of all living things in an area. (3) The temperate zone is found between the arctic and tropical areas and has a moderate temperature range.

54. **(3)** Land biomes exhibit a slow change in vegetation types and are named according to the type which persists for the longest time. INCORRECT CHOICES: (1) The characteristic animals are dependent on the vegetation present. (2) Major climatic changes can destroy the flora and fauna of the region. (4) Major land biomes are found at every temperature and latitude.

55. **(4)** The coastal zone would produce the most in that it is shallow enough to utilize the solar energy for photosynthesis. It contains minerals in the ocean sediments to provide for autotrophs; as the number of autotrophs increases so does the heterotroph population. INCORRECT CHOICES: (1), (2), and (3) All of these choices are areas in which the full solar energy complement cannot be used due to increasing depth.

56. **(1)** Water takes longer to heat up due to the incoming solar energy but retains this heat for a longer period of time than land does. As air masses pass over this heated water and move inland, some of the heat is absorbed and carried to the land mass thereby raising the temperature of the land. INCORRECT CHOICES: (2) The higher humidity tends to advance those autotrophs demanding that condition and the flora flourish. (3) Proximity to salt water moderates the temperature. (4) Soil fertility is not usually enhanced by proximity to a marine biome.

57. **(4)** Traveling south from the Arctic, the frozen tundra would be the first biome encountered; as you reach the equator, the tropical rain forest would be the prevalent biome. INCORRECT CHOICES: (1) A temperate forest would not be found at the Arctic Circle. (2) The taiga or coniferous forest is found south of the tundra. (3) A tropical forest, found at the equator, would not occur immediately below the tundra.

58. **(3)** Nutrient materials and dissolved oxygen pass from the mother's blood to the fetal blood in the placenta, without a direct mixing of the blood. INCORRECT CHOICES: (1) The amnionic sac surrounds the embryo and protects it from shock. (2) There can be absolutely no mixing of maternal and fetal blood. (4) The heart of the fetus is formed early in embryonic development.

59. **(2)** A saprophyte is an organism such as a mold which lives on dead organic matter. INCORRECT CHOICES: (1) A parasite is a symbiont which benefits at the expense of its living host. (3) A herbivore is an animal which eats plants. (4) A predator is an animal which kills and feeds upon living animals.

60. **(4)** The replacement of one community by another until a climax stage is reached is known as ecological succession. INCORRECT CHOICES: (1) Biological control refers to the use of one species to control the numbers or activities of another. (2) Species migration refers to the periodic movements of populations from one geographic location to another. (3) Population fluctuation refers to changes in population size.

61. **(2)** Decomposers break down wastes and dead organisms to simple inorganic materials and make these materials available for reuse. INCORRECT CHOICES: (1) In most natural communities, producers do outnumber consumers. (3) Oxygen supply is maintained by the activity of photosynthetic organisms. (4) Carbon dioxide is used in the process of photosynthesis by green plants.

62. **(1)** Of the four choices, this number contains all abiotic or nonliving factors. INCORRECT CHOICES: (2), (3), and (4) All of these choices contain abiotic factors along with one biotic factor.

63. **(3)** A community includes all the species of organisms in a given location interacting with one another. INCORRECT CHOICES: (1) A population includes all members of a species in a given location. (2) Organisms in the same species are capable of interbreeding and producing fertile offspring. (4) A niche refers to the role an organism plays in a particular habitat; what food and materials it takes from the environment, what it contributes, and how it affects others.

64. **(4)** Athlete's foot, a fungus, is a parasite living on humans. INCORRECT CHOICES: (1) The alga and fungus both benefit from their symbiotic union in the lichen. (2) The nitrogen-fixing bacteria and the clover on which it lives exist in a mutually benefical relationship. (3) The shark is not harmed by the remora eating the shark's excess food.

65. **(3)** A community is built on many different species in a given area and their interactions. INCORRECT CHOICES: (1) A population is the number of the same species inhabiting an area. (2) The biosphere includes all the ecosystems. (4) A biome is a climax community typical of a broad area of climate.

66. **(2)** When two organisms living in the same environment utilize the same limited resource, that is, occupy the same ecological niche, competition between them occur. INCORRECT CHOICES: (1) Reversal of material cycles is not possible because each organism inherits genes which make possible certain metabolic pathways. (3) The populations in a climax community exist in balance with each other and with the environment. (4) If two species compete for the same niche, one, due to a higher rate of reproduction, may eliminate the other which may become extinct.

67. **(4)** A greater number of heterotrophs than autotrophs would mean that the energy flow would soon stop. Therefore the heterotrophs which rely on autotrophs for food would starve. INCORRECT CHOICES: (1) A constant source of energy would ensure the survival of many numbers of autotrophs and thereby sustain the system. (2) The cycling of elements is necessary to keep the balance of the system. (3) This choice is a description of an autotroph.

68. **(2)** Abiotic means no life or the absence of life. Factors which are included in this category are those listed. These factors also are limiting in their affect of organisms. INCORRECT CHOICES: (1) The abiotic factors listed are of great importance to the survivial of the organism. Low rainfall would mean a short water supply and water is the key chemical in all cells, wide temperature variations would mean the difficulty in adaptation would increase, and the shortened periods of daylight would reduce photosynthesis and affect the entire food chain. (3) As stated these factors are abiotic, not biotic. (4) Again these factors are abiotic.

69. **(4)** The organism is the simplest level of biotic organization and as you proceed through the list, you find increasingly more complex relationships. INCORRECT CHOICES: (1) The term population should come before community. (2) Here two terms are out of order; organism and biosphere. (3) Here the term ecosystem should come before biosphere.

70. **(2)** Mutualism is the relationship whereby both organisms benefit. In this example, the nodules of the plant afford a niche for the bacteria which in turn release valuable nitrogen for the plant's growth. INCORRECT CHOICES: (1) Commensalism is the situation where one organism benefits from consuming the unused food of the other organism. The second organism is neither harmed nor receives any benefit. (3) Parasitism is the relationship where the host is harmed by the nutritional relationship of the parasite. (4) Saprophytism is the acquisition of nutrients from dead and decaying organic matter.

71. **(1)** Lichens are usually the first plants to populate an area of bare rock. INCORRECT CHOICES: (2) Grasses are not pioneer organisms. (3) Nitrogen-fixing bacteria live on the roots of plants known as legumes. (4) Hemlock trees require a subsurface of soil.

72. **(3)** Biotic factors are those involving interactions among living organisms, therefore, the number of trees competing for water and light is a biotic factor. INCORRECT CHOICES: (1), (2), and (4) Hours of sunlight, variations in temperature, and quantity of oxygen are nonliving or abiotic factors in the environment.

73. **(1)** Decomposers are organisms which feed upon dead organic matter returning materials to the environment for use by other living things. INCORRECT CHOICES: (2) Herbivores are plant-eating animals and not directly involved in decomposition. (3) Producers are green plants capable of carrying on photosynthesis. (4) A parasite is an organism that lives in or on a host from which it derives its nutrition.

74. **(2)** Insects are the most numerous animals and various species are adapted to eating all the foods which humans eat. INCORRECT CHOICES: (1) Rabbits are herbivores eating many plant species which humans do not eat; also, the total rabbit population is small compared to insect populations. (3) Earthworms eat soil and therefore, do not compete with humans. (4) Rats are important competitors for human food supplies but their numbers are much smaller than those of insect populations.

75. **(3)** A hawk is a meat eater and the word carnivore means meat eater. INCORRECT CHOICES: (1) A parasite feeds off the host organism usually causing it harm. (2) A herbivore eats strictly vegetation. (4) A saprophyte takes its nourishment from dead or decaying organic matter.

76. **(1)** Because there is a loss of energy at each feeding level in a community, the producers contain the greatest amount of energy. INCORRECT CHOICES: (2) Primary consumers eat the producers and use some energy in doing so and for their own metabolism, hence they contain less total energy than the producers. (3) Secondary consumers eat the primary consumers and at this link in the food web more energy is lost. (4) Decomposers break down wastes and dead organisms and they, too, in their metabolism consume energy and give off additional energy as heat.

77. **(2)** DDT enters the food web by being absorbed by forage crops; it then can result in the extinction of various types of animals who feed on them. INCORRECT CHOICES: (1) Although crop production may temporarily increase, the harmful effects of DDT outweigh this benefit. (3) Most insects are not immune to pesticides, such as DDT; those that are, survive and reproduce. (4) DDT is a molecule which becomes part of the material forming the food chain.

78. **(1)** Marine environments are the most stable biomes; they have a relatively constant supply of materials and temperature. INCORRECT CHOICES: (2) In the desert there is a great fluctuation in temperature and rainfall. (3) Deciduous forests are characterized by cold winters and hot summers. (4) In the tundra there can be wide variations in the temperature of the surface strata.

79. **(3)** The roots of trees hold soil particles together and water is held between the soil particles. INCORRECT CHOICES: (1) Photosynthesis in the leaves produces the food supply for the forest community but has relatively little effect on flooding. (2) Transpiration, or the loss of water from leaves, tends to increase the humidity of the air over a forest. (4) Water passes from the roots to the leaves through the xylem ducts.

80. **(2)** More than 70% of the earth's surface is covered by water and the greatest amount of food production (photosynthesis) in the world occurs in the oceans. INCORRECT CHOICES: (1) As altitude increases in the mountains, temperature often becomes the limiting factor in determining the kinds and number of plants which can survive. (3) In the desert, lack of water limits the growth of plants. (4) Tundras are found at high altitudes and in the far north where low temperatures limit plant growth.

81. **(1)** Autotrophic organisms produce oxygen as a result of photosynthesis. INCORRECT CHOICES: (2) Heterotrophs must take in preformed organic molecules as they cannot carry on photosynthesis. (3) Fungi lack chlorophyll and cannot carry on photosynthesis. (4) Carnivores are animals which eat animals.

82. **(4)** Autotrophs capture light energy from the Sun and use it in the synthesis of organic molecules. INCORRECT CHOICES: (1) Fish obtain energy from the food which they eat. (2) Autotrophs must have light energy in order to perform photosynthesis. (3) Water is a raw material for photosynthesis; in the light reactions light energy is used to split water molecules.

83. **(3)** Since there is a loss of energy at each feeding level in a community, the producers contain the greatest amount of energy. INCORRECT CHOICES: (1) Primary consumers feed directly upon green plants; some of the energy in the plants is used by the consumers as they eat, some in their own metabolism, and some is given off as heat. (2) Secondary consumers prey upon primary consumers and the total amount of energy is still further decreased. (4) Herbivores are plant-eating animals.

84. **(1)** In order to produce fertile offspring the two organisms which are mated must be in the same species. INCORRECT CHOICES: (2), (3), and (4) Organisms which share the same environments, or require the same food materials, or have structures for aerobic respiration, need not be in the same species and may not be able to reproduce when mated to each other.

85. **(1)** Since the DDT killed the predators of the red mite, a natural check on the growth of the red mite population was removed. INCORRECT CHOICES: (2) If DDT killed both beneficial and harmful species, then a decrease in both the red mite population and the predator population would be expected. (3) DDT is an insecticide, not a nutrient. (4) Mutations are usually harmful to the organism in which they occur.

86. **(2)** Bacteria are saprophytes which take food from dead or decaying organic matter. INCORRECT CHOICES: (1) A parasite is an organism which lives on or in a living host from which it takes its food. (3) An autotroph is an organism which synthesizes its food by photosynthesis. (4) Multicellular plants contain vascular bundles for the transport of food and water throughout the organism. A bacterium is a protist, a one-celled organism.

87. **(1)** An increase in latitude causes a decrease in the temperature in much the same way as an increase in altitude does. INCORRECT CHOICES: (2) An increase in altitude would result in a decrease of moisture. (3) An increase in altitude would result in an increase in available light. (4) A decrease in longitude would have little effect.

88. **(1)** Approximately 90% of the photosynthesis on this planet occurs in the coastal ocean waters to a depth of about 200 meters where light can penetrate. INCORRECT CHOICES: (2) A desert biome is all the living organisms in and on a land area where there is little rainfall and greatly fluctuating temperatures. (3) A taiga forest exists on a land surface with cold temperatures, a summer growing season, and where coniferous trees predominate. (4) Tundra biomes are found in the land surrounding the Arctic ocean. It is permanently frozen to a depth of a few meters and contains relatively few living organisms.

89. **(4)** Because of the constant supply of nutrients and the stable temperature, the marine biome shows the least changes. INCORRECT CHOICES: (1) A desert shows wide fluctuations in temperature during a 24-hour period. (2) Although a temperate deciduous forest is more stable than some other biomes, there are wide seasonal variations not found in a marine environment. (3) A coniferous forest or taiga has greater temperature fluctuations than a marine biome does.

90. **(4)** By definition, the climax plants are the dominant ones in a stable, self-perpetuating community. They will not be replaced unless there is a major climatic, geologic, or biotic change. INCORRECT CHOICES: (1) (2) Both mosses and grasses are small plants which would appear before the large trees in a forest succession. (3) The pioneer plants would be the first plants in the region.

Posttest

Now that you have finished your review of the subject, you should evaluate your progress. Ideally, you should know enough about the subject to do well on any final examination. However, after taking this Posttest, you will have an indication of any remaining areas in which you may still have a weakness. Go back to the appropriate chapters to review the material. If you are still unclear about the material, ask your teacher, or consult your textbook.
  Good luck!

Directions: Each question is followed by four choices.
Underline the correct chocies.

1. Chloroplasts are cell structures that are located in the
   (1) endoplasmic reticulum   (3) cytoplasm
   (2) cell wall   (4) nucleus

2. Which structure has a similar function in both a human skin cell and a bean leaf cell?
   (1) cell wall   (3) contractile vacuole
   (2) chloroplast   (4) cell membrane

3. Pathways for the transport of materials within a living cell are provided by the
   (1) centrosome   (3) ribosome
   (2) endoplasmic reticulum   (4) cell membrane

4. Which term refers to the chemical substance that aids in the transmission of the impulse through the area indicated by X?

IMPULSE   IMPULSE

(1) neurohumor   (3) neuron
(2) synapse   (4) nerve

5. The function of the setae of the earthworm is to
   (1) provide temporary anchorage in soil
   (2) remove toxic wastes of cell metabolism
   (3) absorb gases used in respiration
   (4) regulate and coordinate muscular contraction

Directions: Base your answers to questions 6 and 7 on this graph and on your knowledge of biology. The graph below depicts changes in the population growth rate of the Kaibab deer.

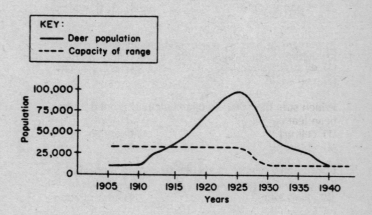

6. About how many deer could the range have supported in 1930 without some of them starving to death?
   (1) 12,000   (3) 50,000
   (2) 35,000   (4) 100,000

7. In which year were the natural predators of the deer most likely being killed off faster than they could reproduce?
   (1) 1905   (3) 1930
   (2) 1920   (4) 1940

8. Compared to Darwin's original theory, the theory of evolution currently accepted by many scientists retains
   (1) none of the points of Darwin's original theory
   (2) components of Darwin's theory and adds modern genetics
   (3) the essentials of Darwin's theory and adds Lamarck's theory
   (4) Darwin's theory with the concept of overproduction deleted

9. When bonded together chemically, deoxyribose, phosphate, and an adenine molecule make up
   (1) a DNA nucleotide   (3) a DNA molecule
   (2) an RNA nucleotide   (4) an RNA molecule

10. A man of blood type AB marries a woman of blood type A. What are the possible blood types of their offspring if the woman's mother was blood type O?
    (1) AB, only   (3) A, B, and O
    (2) A and B, only   (4) A, B, and AB

11. In humans, a primary sperm cell usually gives rise to
    (1) two $2n$ polar bodies   (3) four n polar bodies
    (2) two $2n$ sperm cells   (4) four n sperm cells

12. In an apple blossom, an egg nucleus is formed in the
    (1) stigma   (3) ovule
    (2) stamen   (4) anther

13. Green algae release molecular oxygen as a result of
    (1) aerobic respiration   (3) photosynthesis
    (2) digestion   (4) transpiration

14. In an aerobic cell, active transport is most directly affected by damage to the
    (1) mitochondira   (3) chloroplasts
    (2) ribosomes   (4) centrioles

15. A heterotroph that benefits at the expense of its living host is known as a
    (1) decomposer   (3) parasite
    (2) saprophyte   (4) scavenger

16. A group of organisms within a given area capable of interbreeding and producing fertile offspring under natural conditions is called
    (1) an ecosystem
    (2) a community
    (3) a food web
    (4) a population

17. Which population has risen most rapidly because of the removal of many natural checks and balances?
    (1) grizzly bear
    (2) human
    (3) blue whale
    (4) crocodile

18. According to the heterotroph hypothesis, the earliest heterotrophs must have
    (1) been able to synthesize organic molecules from inorganic compounds
    (2) used oxygen from the atmosphere for respiration
    (3) survived on existing organic molecules in the seas
    (4) been unable to carry on anaerobic respiration

19. If a fossil mammoth were discovered frozen in ice, its cells could be analyzed to determine whether its proteins were similar to those of the modern elephant. This type of investigation is known as comparative
    (1) anatomy
    (2) embryology
    (3) biochemistry
    (4) ecology

20. If bean plant seedlings are germinated in the dark, the seedlings will lack green color. The best explanation for this condition is that
    (1) bean plants are heterotrophic organisms
    (2) bean seedlings lack nitrogen compounds in their cotyledons
    (3) the absence of an environmental factor limits the expression of a genotype
    (4) bean plants cannot break down carbon dioxide to produce oxygen in the dark

21. Each cell in the leaf bud of a plant contains 26 chromosomes. How many chromosomes would be found in each cell of the mature leaf formed from the bud by mitotic cell division?
    (1) 13
    (2) 26
    (3) 28
    (4) 52

Directions: Base your answers to questions 22 through 24 on the pedigree chart below which shows a history of blood types.

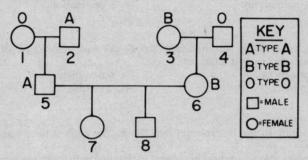

22. The genotype of the individual represented in the chart by 5 is
    (1) $I^a I^a$
    (3) ii
    (2) $I^a i$
    (4) $I^a I^b$

23. Which individuals represented by the chart *must* be homozygous for blood type?
    (1) 1 and 2
    (3) 3 and 4
    (2) 2 and 3
    (4) 1 and 4

24. The blood types of the individuals represented in the chart by 7 and 8 could be
    (1) A or B, only
    (3) A, B, or AB, only
    (2) AB, only
    (4) A, B, AB, or O

25. Identical twins would result from the
    (1) fertilization of two eggs
    (3) same zygote
    (2) eggs from the same ovary
    (4) formation of different zygotes

26. An environmental change which would most likely increase the rate of photosynthesis in a bean plant would be an increase in the
    (1) intensity of green light
    (2) concentration of nitrogen in the air
    (3) concentration of oxygen in the air
    (4) concentration of carbon dioxide in the air

27. Experimental evidence has shown that auxins are responsible for
    (1) differentiation in developing animal embryos
    (2) transport of water down plant stems
    (3) regeneration of animal tissues
    (4) plant growth responses to light and gravity

28. As a direct result of which life process does a plant make a variety of chemical substances such as poisons, drugs, and flavorings?
    (1) digestion   (3) respiration
    (2) excretion   (4) synthesis

29. In Paramecia, most intracellular hydrolysis occurs within structures known as
    (1) ribosomes   (3) mitochondria
    (2) endoplasmic reticula   (4) food vacuoles

30. Which organism possesses an open circulatory system?
    (1) earthworm   (3) Ameba
    (2) grasshopper   (4) Hydra

31. Lymph is an intercellular fluid that originates from
    (1) bile   (3) urine
    (2) plasma   (4) water

32. Which compound is formed in the process of photosynthesis?
    (1) DNA   (3) colchicine
    (2) PGAL   (4) ammonia

33. Which will most probably occur when two organisms in the same habitat occupy the same niche?
    (1) competition   (3) geographic isolation
    (2) mutation   (4) symbiosis

34. Which instrument would provide the most detailed information about the internal structure of a chloroplast?
    (1) a compound light microscope   (3) an electron microscope
    (2) a phase contrast microscope   (4) an interference microscope

35. Which organism is a primary consumer?
    (1) predator   (3) decomposer
    (2) scavenger   (4) herbivore

36. In which area of the Earth would the greatest amount of food normally be produced?
    (1) Atlantic Ocean
    (2) Kaibab Forest
    (3) Sahara Desert
    (4) Rocky Mountains

37. Two varieties of moths existed in a given environment. The table below shows the number of each variety before, shortly after, and long after the construction of a factory in the environment.

	Before	Shortly After	Long After
Green Variety	3,000	2,500	500
Gray Variety	750	2,000	2,800

Which is the most likely explanation for the observed change in the population?
    (1) The change in the environment selected in favor of the gray moth and against the green moth.
    (2) The green moths adapted to the changed environment and became gray.
    (3) The factory output increased the mutation rate at which the gray moth mutated to green.
    (4) Green moths migrated into the area as gray moths migrated out of the area.

38. Which condition would tend to cause a change in the gene frequencies within a population?
    (1) no mutations
    (2) no migration
    (3) large populations
    (4) controlled matings

39. The evolution of antibiotic-resistant strains of bacteria is an illustration of
    (1) natural selection
    (2) use and disuse
    (3) random mating
    (4) geographic isolation

40. Which best describes adaptive radiation?
    (1) Old species always become extinct when new species evolve.
    (2) Species adapt to radioactive substances in their environments.
    (3) Species evolve filling available environmental niches.
    (4) More than one species evolve to fill a single niche.

41. The term "gene pool" refers to
    (1) some of the traits in a given population
    (2) some of the traits in a given species
    (3) the total of all the heritable genes for all the species in a given community
    (4) the total of all the heritable genes for all the traits in a given population

42. If a colorblind man marries a woman who is a carrier for color blindness, it is most probable that
    (1) all of their sons will have normal color vision
    (2) half of their sons will be colorblind
    (3) all of their sons will be colorblind
    (4) none of their children will have normal color vision

43. Two types of RNA molecules are
    (1) uracil and adenine
    (2) messenger RNA and transfer RNA
    (3) cytosine and thymine
    (4) transfer RNA and translocation RNA

44. From which embryonic layer do muscle and bone develop?
    (1) epidermis
    (2) mesoderm
    (3) ectoderm
    (4) endoderm

45. Embryos of both sea and land animals develop in a watery environment. The fluid for the developing land animal is found within the
    (1) umbilical cord
    (2) yolk sac
    (3) amnion
    (4) allantois

46. Phototropism is a plant growth response which most directly depends upon the unequal distribution of
    (1) auxins
    (2) neurohumors
    (3) simple sugars
    (4) mutagenic agents

47. Fungi can absorb food from their environment as a direct result of
    (1) digestion outside the cells
    (2) digestion within the cells
    (3) the action of numerous vacuoles
    (4) the action of large stomates

48. A result of nutrient hydrolysis in animals is the
    (1) conversion of soluble molecules to insoluble molecules
    (2) removal of water from nutrient molecules
    (3) conversion of a large molecule to two or more smaller ones
    (4) conversion of starches to glycogen and cellulose

49. In humans and earthworms, digested foods pass through the intestinal wall and are absorbed into the
    (1) stomach
    (2) Malpighian tubules
    (3) tracheal tubules
    (4) blood

50. In an animal cell, DNA is found in greatest concentration in the
    (1) vacuole
    (2) ribosome
    (3) nucleus
    (4) centrosome

51. The molecular formulas of most carbohydrates have a 2:1 ratio between atoms of
    (1) hydrogen and oxygen
    (2) nitrogen and carbon
    (3) hydrogen and nitrogen
    (4) carbon and oxygen

52. The diagram below represents the field-of-vision of a compound light microscope. What is the approximate length of the cell shown in the diagram?

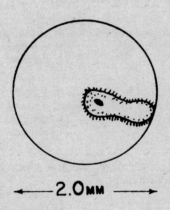

← 2.0 MM →

    (1) 10μ
    (2) 100μ
    (3) 1,000μ
    (4) 2,000μ

53. An abiotic factor which affects the ability of pioneer organisms such as lichens to survive is the
    (1) type of climax vegetation
    (2) species of algae
    (3) type of substratum
    (4) species of bacteria

54. A factor that tends to cause species to change is a
    (1) stable environment
    (2) lack of migration
    (3) recombination of genes
    (4) decrease of mutations

55. At certain stages of development, gill slits may be observed in chicken and human embryos. The presence of gill slits suggests that

(1) humans and chickens may have evolved from a common ancestor
(2) humans evolved from the chicken
(3) the chicken is a fish at one stage of development
(4) humans and chickens developed internal gills useful in adult life

56. In pea plants, reddish-purple seed coat color is dominant over white seed coat color. In a cross between pea plants hybrid for reddish-purple seed coat color, what percent of the offspring would be expected to have white seed coats?
(1) 0%
(2) 25%
(3) 50%
(4) 100%

57. In pea plants, the trait for tall stems is dominant over the trait for short stems. If two heterozygous tall plants are crossed, what percentage of the offspring would be expected to have the same *phenotype* as the parents?
(1) 25%
(2) 50%
(3) 75%
(4) 100%

58. The chromosome number in an egg cell nucleus of a plant is 14. The normal chromosome number in a root epidermal cell of the same plant is
(1) 7
(2) 14
(3) 21
(4) 28

59. Replication of genetic material results in a double-stranded chromosome. An individual strand is known as a
(1) centromere
(2) centriole
(3) centrosome
(4) chromatid

60. Compared to animals that carry on internal fertilization, animals that carry on external fertilization usually
(1) produce fewer eggs
(2) reproduce in water
(3) reproduce by spores
(4) display more parental care

61. In most multicellular animals, meiotic cell division occurs in specialized organs known as
(1) gonads
(2) gametes
(3) kidneys
(4) cytoplasmic organelles

Directions: Base your answers to questions 62 through 66 on the summary word equations below which represent some common biological processes.

(1) Glucose + oxygen $\xrightarrow{\text{enzymes}}$ water + carbon dioxide + energy

(2) Glucose $\xrightarrow{\text{enzymes}}$ ethyl alcohol + carbon dioxide + energy

(3) Water $\xrightarrow[\text{light}]{\text{chlorophyll}}$ hydrogen + oxygen

(4) Hydrogen + carbon dioxide $\xrightarrow{\text{enzymes}}$ phosphoglyceraldehyde (PGAL)

62. Which equation represents a process in animals that releases the greatest amount of energy per molecule of substrate?
    (1) 1   (3) 3
    (2) 2   (4) 4

63. Which equation represents a photochemical process?
    (1) 1   (3) 3
    (2) 2   (4) 4

64. Which equation represents a process that occurs in yeast cells under anaerobic conditions?
    (1) 1   (3) 3
    (2) 2   (4) 4

65. Which equation represents carbon fixation?
    (1) 1   (3) 3
    (2) 2   (4) 4

66. Which equations represent processes that normally occur in most green plants?
    (1) 1 and 4, only   (3) 1, 3, and 4, only
    (2) 3 and 4, only   (4) 1, 2, 3, and 4

67. Which structure is *not* considered to be an endocrine gland?
    (1) thyroid   (3) pancreas
    (2) gall bladder   (4) testis

68. Heterotrophic protists are similar to animals in that both
    (1) convert simple inorganic molecules into complex organic molecules
    (2) synthesize protein molecules from amino acids
    (3) use chlorophyll as a catalyst during photosynthesis
    (4) can perform only anaerobic respiration

Directions: Base your answers to questions 69 and 70 on the graph below which shows a relationship between germination and temperature for four types of spores.

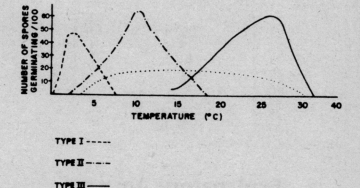

TYPE I -----

TYPE II --·--·-

TYPE III ⎯⎯⎯

TYPE IV ·········

69. The declining rate of germination of type III spores is most likely the result of an increase in
    (1) enzyme deactivation
    (2) dehydration synthesis
    (3) oxidation of glucose
    (4) diffusion of oxygen

70. Which spores germinate over the widest temperature range?
    (1) type I
    (2) type II
    (3) type III
    (4) type IV

Answers

1. 3	16. 4	31. 2	46. 1	61. 1
2. 4	17. 2	32. 2	47. 1	62. 1
3. 2	18. 3	33. 1	48. 3	63. 3
4. 1	19. 3	34. 3	49. 4	64. 2
5. 1	20. 3	35. 4	50. 3	65. 4
6. 1	21. 2	36. 1	51. 1	66. 3
7. 2	22. 2	37. 1	52. 3	67. 2
8. 2	23. 4	38. 4	53. 3	68. 2
9. 1	24. 4	39. 1	54. 3	69. 1
10. 4	25. 3	40. 3	55. 1	70. 4
11. 4	26. 4	41. 4	56. 2	
12. 3	27. 4	42. 2	57. 3	
13. 3	28. 4	43. 2	58. 4	
14. 1	29. 4	44. 2	59. 4	
15. 3	30. 2	45. 3	60. 2	

Explanatory Answers

1. **(3)** The cytoplasm is the site of the chloroplasts. INCORRECT CHOICES: (1) The endoplasmic reticulum is also located in the cytoplasm. (2) The cell wall surrounds plant cells. (4) Chloroplasts are not found in the nucleus.

2. **(4)** The cell membrane is the only structure listed which is common to both plant and animal cells. INCORRECT CHOICES: (1) The cell wall is found only in plant cells. (2) The chloroplasts are found only in plant cells. (3) The contractile vacuole functions to excrete excess water and is not present in the cells listed.

3. **(2)** The intricate canal (endoplasmic reticulum), oftentimes connecting the cell and nuclear membranes, is the means of transport of materials from the external environment to the internal environment. INCORRECT CHOICES: (1) The centrosome plays a role in cellular reproduction. (3) The

ribosomes, again, play a role in protein synthesis. (4) The cell membrane forms an interface between the cytoplasm and the external environment.

4. (1) A neurohumor is a fluid found in the synapse which aids in the transmission of the nerve impulse. INCORRECT CHOICES: (2) The synapse is the space between the terminal branches of a neuron and the dendrites of another. (3) The neuron is a nerve cell. (4) A nerve may be composed of one or more neurons.

5. (1) Setae are pairs of bristles located in each segment; they anchor the earthworm in the soil. INCORRECT CHOICES: (2) Nephridia remove toxic wastes from the earthworm. (3) The moist skin of the earthworm is adapted for gas exchange. (4) The nervous system of the earthworm regulates and coordinates muscular contraction.

6. (1) The capacity of the range in 1930 was about 12,000 deer. INCORRECT CHOICES: (2), (3), (4). All are wrong, since the graph indicates a constant capacity of 35,000 deer from 1905 to about 1925. The question asks about 1930.

7. (2) The years 1910–1925, as depicted on the graph, is a time when the deer population increased, possibly due to a decrease in natural predators. INCORRECT CHOICES: (1) The deer population was constant from 1905 to 1910 indicating a balance in the predator–prey relationship. (3) In 1930, the decrease in the deer population was likely due to overgrazing by the large population in 1925—a lack of food. (4) In 1940 there seems to be a balance in the number of deer and the capacity of the range. The deer population would not have decreased so rapidly if the predators were being killed off at a rapid rate.

8. (2) Darwin's theory was correct although he could not explain the origin of variations. This has been corrected by adding the information gathered through modern genetics. INCORRECT CHOICES: (1) In that Darwin's theory is correct, it has been kept. (3) Lamarck's theory was disproven by Weismann so there would be no need to keep it. (4) The concept of overproduction was based on Malthus's essay and is also true.

9. (1) A DNA nucleotide is composed of three parts, deoxyribose, phosphate, and a nitrogenous base (adenine). INCORRECT CHOICES: (2) In RNA nucleotides, ribose is substituted for deoxyribose. (3) and (4) Both DNA and RNA molecules are composed of nucleotide building blocks.

10. (4) Since the question states that the wife's mother had type O blood we should complete the Punnett squares for: Type AB crossed with AA and

another square: Type *AB* crossed with heterozygous *A*. Completing these we find that blood types *AB*, *A*, and *B* are allowed. INCORRECT CHOICES: (1), (2), and (3) Following the directions for the Punnett squares above it is easy to see why these choices are incorrect.

11. **(4)** Each primary sperm cell undergoes meiosis and gives rise to four monoploid (*n*) sperm cells. INCORRECT CHOICES: (1) and (3) Meiosis of a primary egg cell gives rise to monoploid (*n*) polar bodies as well as one functional egg. (2) As a result of meiosis, sperm cells are monoploid (*n*), not diploid (*2n*).

12. **(3)** In flowers, the female reproductive organs are the pistils; the ovary is part of the pistil and contains ovules within which the egg nuclei are formed. INCORRECT CHOICES: (1) The stigma is the part of the pistil which receives the pollen grains in pollination. (2) The stamen is the male reproductive organ of the flower. (4) The anther is part of the stamen.

13. **(3)** During the light reactions of photosynthesis in green algae, light energy absorbed by chlorophyll is used to "split" water molecules, releasing hydrogen atoms and oxygen gas. INCORRECT CHOICES: (1) Aerobic respiration uses molecular oxygen as a hydrogen acceptor. (2) Digestion breaks large organic molecules into their building blocks. (4) Transpiration refers to the loss of water from the leaves of a plant.

14. **(1)** The function of the mitochondria, "powerhouse of the cell," is to provide energy, in the form of ATP, for such activities as active transport. INCORRECT CHOICES: (2) Ribosomes are involved with the manufacture of protein molecules. (3) Chloroplasts are involved with autotrophic nutrition. (4) Centrioles, found in animal cells, are involved with cellular reproduction.

15. **(3)** A parasite is a symbiotic organism which benefits at the expense of its living host. INCORRECT CHOICES: (1) A decomposer is an organism which breaks down wastes and dead organisms to simpler compounds. (2) A saprophyte is a plant which lives on dead organic matter. (4) A scavenger is an animal which eats organisms that it has not killed.

16. **(4)** A population consists of all members of a species inhabiting a given location; members of a species are capable of interbreeding and producing fertile offspring. INCORRECT CHOICES: (1) The living community and nonliving environment function together as an ecosystem. (2) A community consists of all the populations interacting within a given environment. (3) The many branchings and cross-branchings among the food chains of a community constitute a food web.

17. **(2)** The human population has risen and is still rising very rapidly because of the removal of many natural checks and balances. INCORRECT CHOICES: (1), (3), and (4) The grizzly bear, blue whale, and crocodile are three species threatened with extinction by human activities.

18. **(3)** The earliest heterotrophs used aggregates of organic molecules as "food." INCORRECT CHOICES: (1) Organic molecules were synthesized from inorganic molecules before the evolution of the first heterotrophs. (2) Oxygen did not exist in the environment of the primitive earth. (4) The earliest heterotrophs evolved a form of anaerobic respiration similar to fermentation.

19. **(3)** The tests conducted on the proteins found in the specimens would involve examination using biochemical methods to determine the possible similarities. INCORRECT CHOICES: (1) Protein molecules could not be analysed for similarities using the naked eye. (2) Neither of the specimens studied was an embryo, therefore embryological similarities could not be detected at this time. (4) The biochemical similarities of organisms is most difficult to determine by comparing the environment in which they lived.

20. **(3)** The bean seed's genotype would be green if grown in full light. However, since the bean is grown in the absence of light, it does not exhibit its true genotype. INCORRECT CHOICES: (1) Beans are autotrophic. (2) The question has stated that only one environmental factor, light, was omitted. (4) Carbon dioxide is not broken down to form oxygen in green plants; the oxygen released comes from the water.

21. **(2)** Mitotic division maintains the original chromosome number. Therefore, after mitotic division, there would appear the same number of chromosomes as in the original cell. INCORRECT CHOICES: (1) The number 13 represents one half the complement of chromosomes and implies meiosis. (3) The number 28 would signify the condition of polyploidy. (4) The number 52 is twice the original complement of chromosomes.

22. **(2)** The genotype of individual 5 is $I^a i$. He has type A blood and must have inherited gene i from his mother with type O blood. INCORRECT CHOICES: (1) Individual 5 cannot be homozygous for gene I^a since his mother has type O blood (genotype ii). (3) Individual 5 has type A blood and therefore must have at least one gene I^a. (4) An individual with genotype $I^a I^b$ will have blood type AB.

23. **(4)** Individuals 1 and 4 have type O blood and must therefore be homozygous (ii). INCORRECT CHOICES: (1) Individual 2 could be heterozygous ($I^a i$) for type A. (2) Individuals 2 and 3 could both be heterozygous, ($I^a i$ and $I^b i$ respectively). (3) Individual 3 could be heterozygous ($I^b i$) for type B.

24. **(4)** The offspring of a father who is heterozygous for type A and a mother who is heterozygous for type B could be type A, B, AB, or O.

	I^a	i
I^b	I^aI^b	I^bi
i	I^ai	ii

INCORRECT CHOICES: (1) Children may inherit I^aI^b and have type AB blood, or may inherit ii and have type O. (2) Children who inherit I^ai will have type A; children who inherit I^bi will have type B; children who inherit ii will have type O. (3) Children who inherit ii will have type O blood.

25. **(3)** Identical twins result when a single fertilized egg or zygote splits. INCORRECT CHOICES: (1) Fertilization of two eggs produces fraternal twins. (2) When twins result from two eggs, they do not have the same genetic makeup and are fraternal regardless of the source of the eggs. (4) Identical twins come from the same, not different, zygotes.

26. **(4)** Carbon dioxide and water are the two raw materials for photosynthesis. If more is present, and other favorable conditions exist, photosynthesis would occur more rapidly. INCORRECT CHOICES: (1) Green light is reflected not absorbed by a green plant. (2) Nitrogen is an element which is not involved in photosynthesis. (3) Oxygen is produced during photosynthesis.

27. **(4)** Auxins or plant hormones affect plant growth and may result in greater or less growth depending on the specific hormones and stimulus. INCORRECT CHOICES: (1) Differentiation in developing animal embryos results from the action of animal hormones. (2) Gravity is a major force for the transport of water down a plant. (3) Auxins are hormones found in plants not animals.

28. **(4)** Synthesis, as we have seen previously, is the manufacture of more complex substances, usually organic from simple inorganic elements that are absorbed from the environment. INCORRECT CHOICES: (1) Digestion is the breakdown of complex substances. (2) Excretion is the removal of wastes formed as a result of general metabolism. (3) Respiration is either the incomplete or complete oxidation of glucose to release energy to carry out the life processes.

29. **(4)** Food vacuoles contain the hydrolytic enzymes necessary for digestion. INCORRECT CHOICES: (1) Ribosomes are the sites of protein synthesis. (2) Endoplasmic reticula are channels in the cytoplasm specialized for transport. (3) Mitochondria are organelles in which aerobic respiration takes place.

30. **(2)** In the grasshopper, the circulatory fluid is not contained in vessels but moves freely within the body cavity. INCORRECT CHOICES: (1) Blood is pumped throughout the closed circulatory system of the earthworm by five aortic arches or "hearts." (3) Materials move by cytoplasmic streaming or cyclosis in the one-celled Ameba. (4) The Hydra does not possess a special transport system, most of its cells are in contact with the environment.

31. **(2)** Lymph is part of the plasma fluid which leaves the circulatory vessels and surrounds some body cells. INCORRECT CHOICES: (1) Bile is produced in the liver and is utilized in the emulsification of fat nutrients. (3) Urine is a liquid waste which contains excess salts and water. (4) Water is a waste material formed by aerobic respiration.

32. **(2)** PGAL or phosphoglyceraldehyde is formed as an end product in the dark reaction of photosynthesis. It is formed as a result of the reduction of PGA by $NADH_2$. INCORRECT CHOICES: (1) DNA is the material which codes heredity according to its sequence of nitrogenous bases. (3) Colchicine is a plant extract which inhibits normal mitotic division. (4) Ammonia is a metabolic waste excreted by unicellular water dwellers.

33. **(1)** Occupancy of the same niche by two organisms whether of the same species or not will result in competition for that space. INCORRECT CHOICES: (2) A mutation would be caused by factors affecting the sex cells or gametes of the organism. Competition would not cause mutations. (3) Geographic isolation would prevent competition. (4) Symbiosis refers to a nutritional relationship where two organisms "live together."

34. **(3)** The electron microscope gives the greatest magnification and shows the most detail. INCORRECT CHOICES: (1) The compound light microscope does not have the magnification necessary to show the detailed structure of a chloroplast. (2) and (4) The phase contrast and the interference microscopes use different techniques to see structures without staining them.

35. **(4)** A primary consumer is defined as one that feeds directly from the population of autotrophs. INCORRECT CHOICES: (1) A predator relies on other heterotrophs. (2) A scavenger relies on dead or decaying organic matter and seldom kills another heterotroph. (3) A decomposer derives nutrition from decayed organic matter.

36. **(1)** The greatest amount of food production occurs in the ocean, particularly in coastal waters. INCORRECT CHOICES: (2) The area of the Atlantic Ocean is much greater than that of the Kaibab Forest. (3) Food production in the desert is limited by the lack of water. (4) The area of the Rocky

Mountains is not as great as that of the Atlantic Ocean; furthermore, marine biomes are far more productive than are land biomes.

37. **(1)** The construction of a factory changed the environment in such a way that the gray moths were better able to survive; that is, the environment selected in favor of the gray moth. INCORRECT CHOICES: (2) There is no evidence for assigning purpose to evolutionary change. (3) If gray moths mutated to green, the number of green moths would increase, which is the opposite of what happened. (4) If green moths migrated into the area, the number of green moths would increase, which is the opposite of what happened.

38. **(4)** Controlled matings implies the selection of certain individuals for mating which would tend to increase the frequency of genes carried by these individuals in the population. INCORRECT CHOICES: (1) Gene frequencies are changed if mutations occur. (2) Migration of individuals into or out of the population may alter the gene frequencies. (3) In large populations, chance events are unlikely to affect gene frequencies.

39. **(1)** The use of antibiotics has increased the survival value of resistance to antibiotics among bacteria, and thus there has been a natural selection of these antibiotic-resistant strains. INCORRECT CHOICES: (2) The theory of use and disuse implies that the bacteria were able to adapt to the requirements set by the new environment. (3) Random mating suggests that the chance of mating would not be affected by the ability to resist the antibiotic; if so, there would be no increase in the number of bacteria with this ability. (4) The appearance of antibiotic-resistant strains is not dependent on the geographic isolation of these strains from others.

40. **(3)** Species tend to evolve so as to occupy those environmental conditions which are least utilized in the area where they exist; such a pattern where species evolve filling available environmental niches is called adaptive radiation. INCORRECT CHOICES: (1) When new species evolve, old species may continue to exist if they are adapted to the particular environmental conditions in which they live. (2) Radioactive substances may cause an increase in the mutation rate in organisms in their vicinity. (4) If two different species occupy the same niche, one or the other, due to differences in reproductive rates, will be successful in eliminating the other.

41. **(4)** The gene pool consists of the total of all the heritable genes for all the traits in a given population. INCORRECT CHOICES: (1) The gene pool includes all genes for all traits in the population. (2) The gene pool describes the total of all the genes in a population. (3) Because individuals

belonging to different species cannot interbreed, the gene pools of different species, even in the same community, are separate.

42. **(2)** In completing the sex-linked traits for this problem we find typically that there is a 50% chance of having male offspring, and that one-half of those sons will be affected by colorblindness. The defective colorblindness gene is carried on the X-chromosome, represented below as X^n, and the normal gene as X^N. The son needs only one defective colorblindness gene to be colorblind, while a female needs two.

X^nY = father
X^NX^n = mother

	X^n	Y
X^N	X^NX^n	X^NY
X^n	X^nX^n	X^nY

50% girls
50% boys
25% colorblind boys
25% colorblind girls

INCORRECT CHOICES: (1), (3) and (4) All of these choices, as seen in the Punnett square above, are not correct.

43. **(2)** Two types of RNA are m-RNA, which carries the genetic code to the ribosomes, and t-RNA, which transports amino acids. INCORRECT CHOICES: (1) Uracil and adenine are purine bases found in RNA. (3) Cytosine and thymine are pyrimidine bases found in DNA and RNA. (4) This answer is correct in saying transfer RNA but incorrect in the translocation RNA.

44. **(2)** Mesoderm gives rise to the skeleton, muscles, circulatory system, and the gonads. INCORRECT CHOICES: (1) The epidermis is the outer layer of the skin. (3) The skin and nervous system develop from the ectoderm. (4) The lining of the digestive and respiratory systems as well as the liver and pancreas develop from the endoderm.

45. **(3)** The amnion encloses the amniotic fluid which provides a water environment for the development of certain land animals. INCORRECT CHOICES: (1) The umbilical cord is the structural and functional attachment of the placenta to the embryo in placental mammals. (2) The yolk sac contains blood vessels which make possible the transport of food to the chick embryo. (4) The allantois functions in excretion and respiration in the chick embryo; in the human, the allantois contributes to the development of the placenta.

46. **(1)** Auxins are plant hormones which are destroyed by the presence of light. Therefore, the side of the plant facing the light will have those auxins destroyed and hence growth will be retarded while the other side of the plant will grow in a normal manner giving the plant the tilted look.

INCORRECT CHOICES: (2) Neurohumors are chemicals secreted by organisms having specialized neural systems. Since plants do not have nerve fibers neurohumors would not exist. (3) Simple sugars take part in providing energy for the plant and as a reactant in dehydration synthesis. (4) The mutagenic agents would necessarily affect the nucleus and more specifically the genes on the chromosomes leading to variations in offspring.

47. **(1)** Fungi appear ecologically to be decomposers and as such practice extracellular digestion followed by absorption. INCORRECT CHOICES: (2) Intracellular digestion is an adaptation which fungi do not possess. (3) Vacuoles are sites of intracellular digestion. (4) Stomates function in transpiration of water and the diffusion of gases into a leaf's internal tissues.

48. **(3)** Nutrient hydrolysis (digestion) involves the conversion of a large molecule to two or more smaller ones. INCORRECT CHOICES: (1) Hydrolysis changes large and/or insoluble food molecules to small, soluble ones. (2) Hydrolysis involves the breakdown of large organic molecules with the addition of water. (4) Starches, glycogen, and cellulose are polysaccharides and their hydrolysis results in the production of simple sugars.

49. **(4)** Blood in humans and earthworms constitutes a transport medium through which nutrients are transported to the cells. INCORRECT CHOICES: (1) The stomach in humans digests proteins and stores food. The earthworm does not have a stomach. (2) Malpighian tubules are used in grasshoppers to concentrate products of deamination for excretion. (3) Tracheal tubules are used by the grasshopper in respiration.

50. **(3)** The nucleus is the control center of the cell and the site for chromosome location. INCORRECT CHOICES: (1) A vacuole is the site for intracellular digestion, excretion of excess water, and other cell processes. (2) The ribosome is the site of protein synthesis. (4) The centrosome plays a role in the division of the animal cell.

51. **(1)** Hydrogen and oxygen are found in a 2:1 ratio in the molecules of most carbohydrates. INCORRECT CHOICES: (2) and (3) Carbohydrates do not contain nitrogen. (4) Carbon and oxygen are always found in carbohydrate molecules, but not necessarily in a 2:1 ratio.

52. **(3)** The length of the specimen is half the diameter of the field; there are 1,000 microns (μ) in 1 millimeter. INCORRECT CHOICES: (1) This represents 1/100 of the length of the specimen. (2) This represents 1/10 of the length of the specimen. (4) This represents the entire length of the diameter of the field.

53. **(3)** The lichen grows on barren rock. It secretes acid which attacks the rock and forms bits of soil. The substratum on which it lives is an abiotic factor. INCORRECT CHOICES: (1) Climax vegetation is a biotic community of plants which is stable and self-perpetuating. Its populations exist in balance with each other and with the environment. (2) A species of an algae would be a biotic or living factor. (4) A species of bacteria would be a biotic or living factor.

54. **(3)** Recombination of genes occurring in the crossing-over in meiosis gives variety to the genetic character of the offspring. INCORRECT CHOICES: (1), (2), and (4) All of these choices would maintain the genetic information and submerge any change.

55. **(1)** Similarities in embryonic structures of chickens and humans suggests common ancestry. INCORRECT CHOICES: (2) The presence of gill slits in both chicken and human embryos suggests a common ancestor, but not that one evolved from the other. (3) The presence of gill slits does not make the chicken a fish. (4) Gill slits are present in the embryos only in chickens and humans and do not develop into gills in the adults.

56. **(2)** According to the laws of genetics this cross would typically produce a ratio of 1 pure dominant: 2 hybrid dominant: 1 recessive. INCORRECT CHOICES: (1), (3), and (4) These answers do not adhere to the ratio for crossing two hybrid organisms.

57. **(3)** Seventy-five percent of the offspring of two heterozygous tall plants would be expected to be tall, the same phenotype as the parents.

	T	*t*
T	TT	Tt
t	Tt	tt

Both *TT* and *Tt* are phenotypically tall; therefore, ¾ or 75% are tall.

INCORRECT CHOICES: (1) If only the homozygous dominant individual, *TT*, were tall, then 25% would be correct. (2) Fifty percent of the offspring will have the same genotype, *Tt*, as the parents. (4) The homozygous recessive individual, *tt*, will be short.

58. **(4)** Body cells have double the number of chromosomes found in sex cells. INCORRECT CHOICES: (1) The number of chromosomes in a root cell is twice, not half, that found in the egg cell. (2) Egg or sperm cells have half the chromosome number found in other plant cells. (3) The normal number of chromosomes in a root cell is double, not triple, that found in the sex cells.

59. **(4)** A chromatid represents a duplicate of the original chromosome. The pair of chromatids composes the entire chromosome. INCORRECT CHOICES: (1) The centromere connects the chromatids. (2) The centriole plays a role in the coming mitotic division. (3) The centrosome is found mainly in animal cells and contains the centriole which functions in cell division.

60. **(2)** External fertilization is commonly found in animals that exist at least for a part of their life in water. They do not possess special internal organs for internal fertilization and rely on the water for a medium for fertilization. INCORRECT CHOICES: (1) By contrast, organisms that practice external fertilization, must produce many more eggs to ensure that a few will survive the harsh environmental conditions and predation. (3) Sporulation is practiced by organisms as a form of asexual reproduction. The question is dealing with sexual reproduction. (4) Less parental care is seen in that there are no direct ties to the offspring while the offspring are developing. Most often the parent abandons the eggs after fertilization.

61. **(1)** Meiotic cell division is a reduction in the number of chromosomes to produce gametes. This specialized process needs specialized organs called gonads. INCORRECT CHOICES: (2) The gametes are the end result of meiosis, not the structures in which the process occurs. (3) The kidneys function in the removal of metabolic wastes and reabsorption of essential water and minerals. (4) Cytoplasmic organelles encompass a wide range of structures which do not take part in meiosis.

62. **(1)** Equation 1 represents aerobic respiration which results in the formation of 38 ATP molecules. INCORRECT CHOICES: (2) Equation 2 describes anaerobic respiration which produces a net yield of 2 ATP. (3) Equation 3 represents the light phase of photosynthesis. (4) Equation 4 depicts the dark phase of photosynthesis.

63. **(3)** Light energy is used to split water in the photochemical step of photosynthesis. INCORRECT CHOICES: (1), (2), and (4) The other equations are chemical reactions; they are not *photo*chemical because they do not require light energy.

64. **(2)** Yeast cells produce ethyl alcohol and carbon dioxide during the process of anaerobic respiration. INCORRECT CHOICES: (1) Yeast cells are not capable of aerobic respiration. (3) and (4) Yeast cells are *not* green plants and cannot carry on photosynthesis.

65. **(4)** CO_2 and H_2 atoms combine chemically during the carbon fixation or dark phase of photosynthesis. INCORRECT CHOICES: (1) or (2) Carbon fixation does not occur during aerobic or anaerobic respiration. (3) CO_2 is not incorporated into organic molecules in the light phase of photosynthesis.

66. **(3)** Aerobic respiration and the light and dark phases of photosynthesis take place in green plants. INCORRECT CHOICES: (1) In addition to 1 and 4, the light phase of photosynthesis takes place in green plants. (2) All green plants carry on aerobic respiration. (4) Green plants do not obtain energy through anaerobic respiration.

67. **(2)** The gall bladder is associated with the liver and stores bile produced in the liver. Through the bile duct the bile travels to the small intestine to emulsify fats. INCORRECT CHOICES: (1), (3), and (4) All of these choices are endocrine glands in that they secrete chemicals directly into the bloodstream.

68. **(2)** All organisms, including heterotrophic protists and animals, synthesize protein molecules from amino acids. INCORRECT CHOICES: (1) Autotrophic organisms convert simple inorganic molecules into complex organic molecules. (3) Green plants and algae use chlorophyll to absorb light enengy during photosynthesis. (4) Many heterotrophic protists and all animals are aerobic organisms.

69. **(1)** We find that the activity of enzymes decreases at specific temperatures. In the type III spores that deactivation temperature must be 25°C. INCORRECT CHOICES: (2) Dehydration synthesis has nothing to do with germination. (2) Oxidation of glucose does not occur in germination but in respiration. (4) The diffusion of oxygen might occur across the cell membrane.

70. **(4)** In examining the graph we find indeed that the type IV spores germinate from a low temperature of 4°C to a high temperature of 30°C. INCORRECT CHOICES: (1) Type I spores germinate from 0°C to 8°C. (2) Type II spores germinate from 3°C to 19°C. (3) Type III spores germinate from 14° C to 31°C.

Posttest
Topic Analysis Key

If you had these answers wrong	Read these chapters
1, 2, 3, 14, 32, 34, 48, 50, 51, 52, 69, 70	The Study of Life
4, 5, 29, 30, 31, 49, 62, 67, 68	Maintenance in Animals
13, 26, 27, 28, 46, 47, 63, 64, 65, 66	Maintenance in Plants
11, 12, 21, 25, 44, 45, 58, 59, 60, 61	Reproduction and Development
9, 10, 20, 22, 23, 24, 42, 43, 56, 57	Genetics
8, 18, 19, 36, 37, 38, 39, 40, 41, 54, 55	Evolution and Diversity
6, 7, 15, 16, 17, 33, 34, 35, 53	Plants and Animals in Their Environment

Biology Regents Examination
June 1982

Part I
Answer all 60 questions in this part.   [70]

Directions (1–60): For *each* statement or question, select the word or expression that, of those given, best completes the statement or answers the question and underline it.

1. A characteristic of all known living organisms is that they
   (1) require oxygen for respiration
   (2) originate from preexisting life
   (3) have complex nervous systems
   (4) carry on heterotrophic nutrition

2. Control of all physiological activities of an organism is necessary to maintain that organism's stability in its environment. This life activity is known as
   (1) nutrition   (3) transport
   (2) respiration   (4) regulation

3. Which cell organelles are the sites of aerobic cellular respiration in both plant and animal cells?
   (1) mitochondria   (3) chloroplasts
   (2) centrosomes   (4) nuclei

4. The diameter of the field of vision of a compound light microscope is 1.5 millimeters. This may also be expressed as
   (1) 15 microns   (3) 1,500 microns
   (2) 150 microns   (4) 15,000 microns

5. An organic compound that has hydrogen and oxygen in a 2:1 ratio would belong to the group of compounds known as
   (1) lipids
   (2) fatty acids
   (3) proteins
   (4) carbohydrates

6. Which inorganic substance found in living matter aids in the diffusion of gases through a cell membrane?
   (1) water
   (2) salt
   (3) phosphorus
   (4) iron

7. An earthworm that has partially entered its burrow can be surprisingly difficult to pull from the ground. This is due primarily to the earthworm's
   (1) chitinous outer covering and legs
   (2) bristle-like setae and muscles
   (3) powerful ventral suckers and claws
   (4) grasping mouth parts and scales

8. Because they aid in the regulation of body processes, neurohumors are most similar to
   (1) hormones
   (2) chitin
   (3) urine
   (4) pigments

9. Wastes that may be excreted from human liver cells include
   (1) water, oxygen, and mineral salts
   (2) water, carbon dioxide, and urea
   (3) hormones, urea, and carbon dioxide
   (4) hormones, oxygen, and water

10. Carbon dioxide released from the interior cells of a grasshopper is transported to the atmosphere through the
    (1) Malpighian tubules
    (2) tracheae
    (3) contractile vacuoles
    (4) lungs

11. Vigorous activity of human voluntary muscle tissues may result in the production of lactic acid. Insufficient amounts of which gas would result in the buildup of lactic acid in muscle cells?
    (1) carbon dioxide
    (2) nitrogen
    (3) oxygen
    (4) hydrogen

12. In animal cells, the energy to convert ADP to ATP comes directly from
    (1) hormones
    (2) sunlight
    (3) organic molecules
    (4) inorganic molecules

13. Which organism has an internal, closed circulatory system which brings materials from the external environment into contact with its cells?
    (1) Ameba   (3) Hydra
    (2) Paramecium   (4) earthworm

14. There is a higher concentration of mineral salts within the body of a Paramecium than in its external water environment. This higher concentration is maintained as a result of the action of
    (1) pinocytosis   (3) CO_2
    (2) cyclosis   (4) ATP

15. Which are produced as a result of the mechanical digestion of a piece of meat?
    (1) amino acids   (3) smaller meat particles
    (2) fatty acids   (4) larger glycerol molecules

16. The complete hydrolysis of carbohydrates results directly in the production of
    (1) glycogen   (3) carbon dioxide
    (2) urea molecules   (4) simple sugars

17. Nutrients are reduced to soluble form within the food vacuoles of the
    (1) grasshopper and earthworm   (3) earthworm and human
    (2) Ameba and Paramecium   (4) Hydra and grasshopper

18. Which locomotive structures are found in some protozoa?
    (1) muscles   (3) cilia
    (2) tentacles   (4) setae

19. Which animal has a ventral nerve cord?
    (1) grasshopper   (3) Hydra
    (2) Ameba   (4) human

20. The two systems that directly control homeostasis in most animals are the
    (1) nervous and endocrine systems
    (2) endocrine and excretory systems
    (3) nervous and locomotive systems
    (4) excretory and locomotive systems

21. Ribosomes supply the cell with complex proteins needed for maintenance and repair by the process of
    (1) oxidation   (3) dehydration synthesis
    (2) digestion   (4) enzymatic hydrolysis

22. Which part of a plant is specialized for anchorage and absorption?
    (1) leaf   (3) stem
    (2) flower   (4) root

23. Removing the tip of the stem of a young plant will most directly interfere with the production of
    (1) sugars   (3) carbon dioxide
    (2) auxins   (4) oxygen

24. Rose oil pressed from petals is used to make perfume. The rose oil is present in plant cells as a result of
    (1) absorption from the Sun
    (2) absorption from the soil
    (3) synthesis from simpler compounds
    (4) synthesis from chlorophyll molecules

25. Nitrogenous compounds may be used by plants in the synthesis of
    (1) glucose   (3) proteins
    (2) waxes   (4) starch

26. In plants, the molecular oxygen concentration of a leaf cell usually increases during the process of
    (1) aerobic respiration   (3) transpiration
    (2) photosynthesis   (4) capillary action

27. Which structures transport food downward in a geranium plant?
    (1) phloem vessels   (3) root hairs
    (2) xylem tubes   (4) chloroplasts

28. In a maple tree, the enzymatic hydrolysis of starches, lipids, and proteins occurs
    (1) extracellularly, only
    (2) intracellularly, only
    (3) both extracellularly and intracellularly
    (4) neither extracellularly nor intracellularly

29. A characteristic of animals that makes them similar to heterotrophic plants is that animals
    (1) obtain preformed organic molecules from other organisms
    (2) need to live in a sunny environment
    (3) are sessile for most of their lives
    (4) use energy to manufacture organic compounds from inorganic compounds

30. One bean plant is illuminated with green light and another bean plant of similar size and leaf area is illuminated with blue light. If all other conditions are identical, how will the photosynthetic rates of the plants most probably compare?
    (1) Neither plant will carry on photosynthesis.
    (2) Photosynthesis will occur at the same rate in both plants.
    (3) The plant under green light will carry on photosynthesis at a greater rate than the one under blue light.
    (4) The plant under blue light will carry on photosynthesis at a greater rate than the one under green light.

31. The diagram below illustrates which type of reproduction?

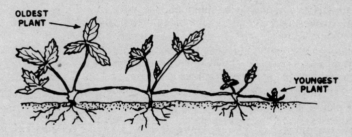

    (1) cleavage   (3) zygote formation
    (2) fission   (4) vegetative propagation

32. Only one member of each pair of homologous chromosomes is normally found in a
    (1) zygote   (3) gamete
    (2) multicellular embryo   (4) cheek cell

33. The production of sperm nuclei in plants occurs in cells from the
    (1) anther   (3) pistil
    (2) stigma   (4) ovary

34. In human males, the maximum number of functional sperm cells that is normally produced from each primary sex cell is
    (1) one   (3) three
    (2) two   (4) four

35. The diagram below represents processes in the reproduction of a honeybee.

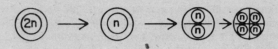

primary sex cell   egg cell   growth of male bee

As indicated in this diagram, which processes produce a male honeybee?
(1) meiosis and budding
(2) meiosis and parthenogenesis
(3) fertilization and cleavage
(4) fertilization and parthenogenesis

36. Each strand of a double-stranded chromosome is known as a
(1) centromere   (3) chromatid
(2) homologue   (4) tetrad

37. Which event occurs in the cytoplasmic division of plant cells but *not* in the cytoplasmic division of animal cells?
(1) cell plate formation   (3) chromosome replication
(2) centromere replication   (4) centriole formation

38. When a cell with 24 chromosomes divides by mitotic cell division, the resulting daughter cells will each have a maximum chromosome number of
(1) 6   (3) 24
(2) 12   (4) 48

39. Organisms which contain both functional male and female gonads are known as
(1) hybrids   (3) phagocytes
(2) hermaphrodites   (4) parasites

40. DNA and RNA molecules are similar in that they both contain
(1) nucleotides   (3) deoxyribose sugars
(2) a double helix   (4) thymine

41. Which is a form of vegetative reproduction used to propagate desirable varieties of plants quickly?
(1) hybridization   (3) fertilization
(2) pollination   (4) grafting

42. In many humans, exposing the skin to sunlight over prolonged periods of time results in the production of more pigment by the skin cells (tanning). This change in skin color provides evidence that

(1) ultraviolet light can cause mutations
(2) gene action can be influenced by the environment
(3) the inheritance of skin color is an acquired characteristic
(4) albinism is a recessive characteristic

43. Which terms best describe most mutations?
(1) dominant and disadvantageous to the organism
(2) recessive and disadvantageous to the organism
(3) recessive and advantageous to the organism
(4) dominant and advantageous to the organism

44. A pair of chromosomes fail to separate during meiosis, producing a gamete with an extra chromosome. This process is known as
(1) crossing-over
(2) polyploidy
(3) nondisjunction
(4) recombination

45. Three brothers have blood types A, B, and O. What are the chances that the parents of these three will produce a fourth child whose blood type is AB?
(1) 0%
(2) 25%
(3) 50%
(4) 100%

46. A colorblind woman marries a man who has normal color vision. What are their chances of having a colorblind daughter?
(1) 0%
(2) 25%
(3) 75%
(4) 100%

47. A student crossed wrinkled-seeded (rr) pea plants with round-seeded (RR) pea plants. Only round seeds were produced in the resulting plants. This illustrates the principle of
(1) independent assortment
(2) segregation
(3) dominance
(4) incomplete dominance

48. The modern classification system is based on structural similarities and
(1) evolutionary relationships
(2) habitat similarities
(3) geographic distribution
(4) Mendelian principles

49. In peas, flowers located along the stem (axial) are dominant to flowers located at the end of the stem (terminal). Let A represent the allele for axial flowers and a represent the allele for terminal flowers. When plants with axial flowers are crossed with plants having terminal flowers, all of the offspring have axial flowers. In this cross the genotypes of the parent plants are most likely
(1) aa × aa
(2) Aa × Aa
(3) AA × Aa
(4) AA × aa

50. In an ecological succession in New York State, lichens growing on bare rock are considered to be
    (1) climax organisms
    (2) pioneer organisms
    (3) primary consumers
    (4) decomposers

51. The pig has four toes on each foot. Two of the toes are very small and do not have a major function in walking. Lamarck probably would have explained the reduced size of the two small toes by his evolutionary theory of
    (1) natural selection
    (2) mutation
    (3) use and disuse
    (4) synapsis

52. The study of mutations is important to the modern theory of evolution because it helps to explain
    (1) differentiation in embryonic development
    (2) stability of gene pool frequencies
    (3) the extinction of the dinosaurs
    (4) the appearance of variations in organisms

53. Many related organisms are found to have the same enzymes and hormones. This suggests that
    (1) enzymes work only on specific substrates
    (2) enzymes act as catalysts in biochemical reactions
    (3) organisms living in the same environment require identical enzymes
    (4) these organisms may share a common ancestry

54. The fossil record may be used as evidence for organic evolution if it is assumed that
    (1) in undisturbed rock, the oldest fossils are in the lowest layers
    (2) fossils in different layers existed at the same time
    (3) all fossils have homologous structures
    (4) all fossils filled the same niche

55. At times hyenas will feed on the remains of animals they, themselves, have not killed. At other times they will kill other animals for food. Based on their feeding habits, hyenas are best described as
    (1) herbivores and parasites
    (2) herbivores and predators
    (3) scavengers and parasites
    (4) scavengers and predators

56. Which is true of most producer organisms?
    (1) They are parasitic.
    (2) They contain chlorophyll.
    (3) They are eaten by carnivores.
    (4) They liberate nitrogen.

57. Which is an abiotic factor in the environment?
   (1) water
   (2) earthworm
   (3) fungus
   (4) human

Directions: Base your answers to questions 58 through 60 on the food chain represented below and on your knowledge of biology.

rosebush → aphid → ladybird beetle → spider → toad → snake

58. Which organism in the food chain can transform light energy into chemical energy?
   (1) spider
   (2) ladybird beetle
   (3) rosebush
   (4) snake

59. At which stage in the food chain will the population with the *smallest* number of animals probably be found?
   (1) spider
   (2) aphid
   (3) ladybird beetle
   (4) snake

60. Which organism in this food chain is herbivorous?
   (1) rosebush
   (2) aphid
   (3) ladybird beetle
   (4) toad

Part II

This part consists of six groups, each containing ten questions. Choose three of these six groups. Be sure that you answer all ten questions in each group chosen. Underline the answers to these questions.

Group 1—Biochemistry

If you choose this group, be sure to answer questions 61–70.

Directions (61–64): For *each* of questions 61 through 64, select the structural formula, *chosen from those below*, which best answers that question, then underline its *number.*

(1)

$$\begin{array}{c} H \\ \diagdown \\ H \end{array} N - \underset{\underset{H}{|}}{\overset{\overset{R}{|}}{C}} - \overset{\overset{O}{\parallel}}{C} \diagdown OH$$

(3) O=C=O

(2)

$$H - \underset{\underset{H}{|}}{\overset{\overset{H}{|}}{C}} - \underset{\underset{H}{|}}{\overset{\overset{H}{|}}{C}} \cdots \underset{\underset{H}{|}}{\overset{\overset{H}{|}}{C}} - \underset{\underset{H}{|}}{\overset{\overset{H}{|}}{C}} - \underset{\underset{H}{|}}{\overset{\overset{H}{|}}{C}} - \overset{\overset{O}{\parallel}}{C} \diagdown OH$$

(4) [ring structure: CH_2OH group with carbon ring containing O, OH, H substituents]

(5) [two joined ring structures (CH_2OH groups) linked by an O bridge — a disaccharide]

61. Which is a structural formula for a component of a fat?

62. Which formula represents a substance formed as a direct result of a dehydration synthesis?

63. Which formula represents a component of all proteins?

64. Which formula represents a monosaccharide?

65. An organic compound formed in the dark reactions of photosynthesis is
    (1) chlorophyll
    (2) oxygen
    (3) H_2O
    (4) PGAL

66. Digestive enzymes which hydrolyze molecules of fat into fatty acid and glycerol molecules are known as
    (1) proteases
    (2) lipases
    (3) maltases
    (4) vitamins

67. In certain bacteria and yeasts, under anaerobic conditions, the oxidation of glucose leads to the production of
    (1) ethyl alcohol
    (2) complex sugars
    (3) oxygen
    (4) starches

Directions: Base your answers to questions 68 through 70 on the graph below and on your knowledge of biology. The graph represents the rate of enzyme action when different concentrations of enzyme are added to a system with a fixed amount of substrate.

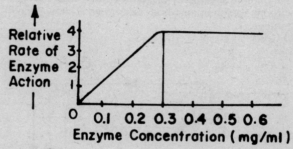

68. At which enzyme concentration does all of the available substrate react with the enzyme?
    (1) 0.1 mg/ml
    (2) 0.2 mg/ml
    (3) 0.3 mg/ml
    (4) 0.05 mg/ml

Note that questions 69 and 70 have only three choices.

69. When the enzyme concentration is increased from 0.5 mg/ml to 0.6 mg/ml, the rate of enzyme action
    (1) decreases
    (2) increases
    (3) remains the same

70. If more substrate is added to the system at an enzyme concentration of 0.4 mg/ml, the rate of the reaction would most likely
    (1) decrease
    (2) increase
    (3) remain the same

Group 2—Human Physiology

If you choose this group, be sure to answer questions 71–80.

Directions: Base your answers to questions 71 through 73 on the schematic diagram below of blood flow throughout the human body and on your knowledge of biology.

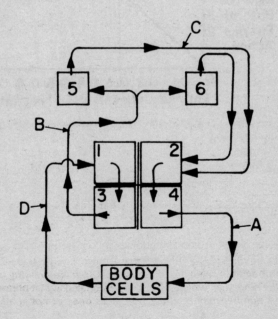

71. Which chambers of the heart contain blood which has the highest concentration of oxygen?
    (1) 1 and 2
    (2) 2 and 4
    (3) 3 and 4
    (4) 1 and 3

72. Which blood vessels contain blood with the lowest concentration of oxygen?
    (1) A and D
    (2) B and C
    (3) C and A
    (4) D and B

73. Microscopic structural units known as alveoli are located in structures
    (1) 1 and 3
    (2) B and C
    (3) 5 and 6
    (4) D and A

74. Which best illustrates the pathway of an impulse in a reflex arc?
    (1) effector → motor neuron → associative neuron → sensory neuron → receptor
    (2) receptor → motor neuron → associative neuron → sensory neuron → effector
    (3) receptor → sensory neuron → associative neuron → motor neuron → effector
    (4) effector → sensory neuron → associative neuron → motor neuron → receptor

75. A person was admitted to the hospital with abnormally high blood sugar and an abnormally high sugar content in his urine. Which gland most likely caused this condition by secreting lower than normal amounts of its hormone?
    (1) pancreas
    (2) parathyroid
    (3) salivary
    (4) thyroid

76. In the human elbow joint, the bone of the upper arm is connected to the bones of the lower arm by flexible connective tissue known as
    (1) tendons
    (2) ligaments
    (3) muscles
    (4) neurons

Directions (77–80): For each phrase in questions 77 through 80 select the organ *chosen from the drawing on the next page*, which is most closely related to that phrase. [A number may be used more than once or not at all.]

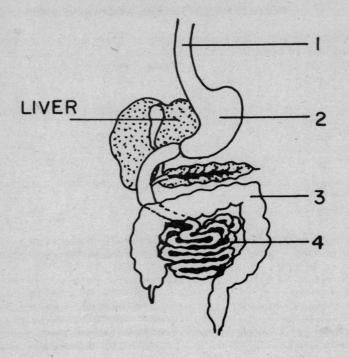

LIVER

77. Where feces are formed

78. Where protein digestion begins

79. Where lipid digestion is completed

80. Contains gastric glands

Group 3—Reproduction and Development

If you choose this group, be sure to answer questions 81–90.

81. In plants, which is a function of the seed parts known as cotyledons?
    (1) formation of flowers
    (2) development of stems
    (3) production of roots
    (4) storage of food

82. The factors necessary for maple seed germination are moisture, proper temperature, and
    (1) oxygen
    (2) soil
    (3) chlorophyll
    (4) darkness

83. The human male's testes are located in an outpocketing of the body wall known as the scrotum. An advantage of this adaptation is that
    (1) the testes are better protected in the scrotum than in the body cavity
    (2) sperm production requires contact with atmospheric air
    (3) a temperature lower than body temperature is best for sperm production and storage
    (4) the sperm cells can enter the urethra directly from the testes

84. In a developing embryo, the process most closely associated with the differentiation of cells is
    (1) gastrulation
    (2) menstruation
    (3) ovulation
    (4) fertilization

Directions (85–87): For each of questions 85 through 87, select the structure in a developing chicken egg, *chosen from the list below,* that best answers the question.

Developing Egg Structures
(1) Amnion
(2) Allantois
(3) Chorion
(4) Yolk sac

85. Which structure most directly protects the embryo from shocks?

86. Which structure lines the shell and surrounds the other membranes?

87. In which structure are embryonic wastes stored?

Directions: Base your answers to questions 88 through 90 on the diagram on the next page which represents a cross section of a part of the human female reproduction system and on your knowledge of biology.

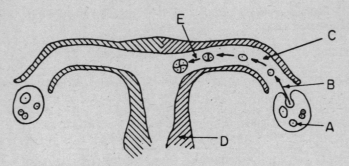

88. Which structure is prepared for implantation of a fertilized egg as a result of the action of reproductive hormones?
    (1) A   (3) C
    (2) B   (4) D

89. Within which structure does fertilization normally occur?
    (1) A   (3) C
    (2) B   (4) D

90. Which represents the process of ovulation?
    (1) A   (3) C
    (2) B   (4) E

Group 4—Modern Genetics

   If you choose this group, be sure to answer questions 91–100.

91. A change which affects the base sequence in an organism's DNA by the addition, deletion, or substitution of a single base is known as
    (1) DNA replication   (3) chromosomal mutation
    (2) gene mutation   (4) independent assortment

92. The replication of a double-stranded DNA molecule begins when the strands separate at the
    (1) phosphate bonds   (3) deoxyribose molecules
    (2) ribose molecules   (4) hydrogen bonds

93. Which nitrogenous bases tend to pair with each other in a double-stranded molecule of DNA?

(1) adenine-uracil
(2) thymine-adenine
(3) cytosine-thymine
(4) guanine-adenine

94. Within which organelles have genes been found?
(1) food vacuoles
(2) contractile vacuoles
(3) mitochondria
(4) cell walls

95. The function of transfer RNA molecules is to
(1) transport amino acids to messenger RNA
(2) transport amino acids to DNA in the nucleus
(3) synthesize more transfer RNA molecules
(4) provide a template for the synthesis of messenger RNA

Directions: Base your answers to questions 96 through 100 on the diagram below, which represents a portion of a messenger RNA molecule associated with a ribosome, and on your knowledge of biology.

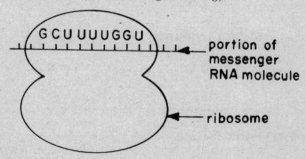

96. The presence of which nitrogen base indicates that the molecule associated with the ribosome is RNA?
(1) guanine
(2) cytosine
(3) uracil
(4) adenine

97. The sequence of nucleotides on the RNA molecule was determined by the
(1) sequence of nucleotides on transfer RNA molecules
(2) base sequence of the original DNA molecule that served as the template
(3) sequence of amino acids that will be linked together to form a polypeptide chain
(4) base sequence of the original messenger RNA molecule that served as a template

98. The messenger RNA genetic codes for 3 different amino acids are:

U-U-U = phenylalanine
G-C-U = alanine
G-G-U = glycine

Using this information, the strip of messenger RNA shown in the illustration would result in an amino acid sequence consisting of
    (1) phenylalanine-alanine-glycine   (3) alanine-glycine-phenylalanine
    (2) alanine-glycine-glycine   (4) alanine-phenylalanine-glycine

99. The association between the ribosome and the messenger RNA molecule occurs in the
    (1) cytoplasm   (3) nucleolus
    (2) centrosome   (4) nucleus

100. The substance being synthesized in the cell is most likely a
    (1) fat   (3) polypeptide
    (2) vitamin   (4) carbohydrate

Group 5—Modern Evolution

If you choose this group, be sure to answer questions 101–110.

101. Which statement concerning living organisms is *least* in agreement with the modern concept of evolution?
    (1) They have nucleic acids as their genetic material.
    (2) They have similar chemical reactions controlled by enzymes.
    (3) They consist of one or more cells.
    (4) They are grouped into species which are unchanging.

102. According to the heterotroph hypothesis, the first organisms were probably
    (1) aerobic heterotrophs   (3) aerobic autotrophs
    (2) anaerobic heterotrophs   (4) anaerobic autotrophs

103. All the genes in a given population which can be inherited constitute a
    (1) gene pool   (3) gene frequency
    (2) genotype   (4) phenotype

104. Several species of milkweed are growing in the same area. All are capable of hybridization but none ever do because the surface of the stigma of each species accepts only a certain shape pollen grain.
This paragraph best describes an evolutionary process known as

(1) artificial selection   (3) survival of the fittest
(2) reproductive isolation   (4) overproduction

105. In a large population located in a constant environment, the following conditions exist: random mating, no migration, and no mutations. Which will most probably occur within the population?
(1) The gene frequencies will remain stable.
(2) The dominant gene frequencies will increase.
(3) The recessive gene frequencies will increase.
(4) All gene frequencies will decrease.

Directions: Base your answers to questions 106 and 107 on the graph and information below and on your knowledge of biology.

Scientists studying a moth population in a woods in New York State recorded the distribution of moth wing color as shown in the graph below. The woods contained trees whose bark color was predominantly brown.

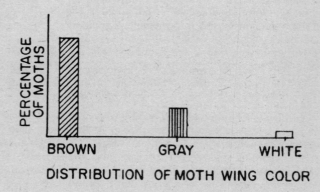

DISTRIBUTION OF MOTH WING COLOR

106. A fungus infection affected nearly all trees in the woods so that the coloration of the tree bark was changed to a gray-white color. Which graph shows the most probable results that would occur in the distribution of wing coloration in this moth population after a long period of time?

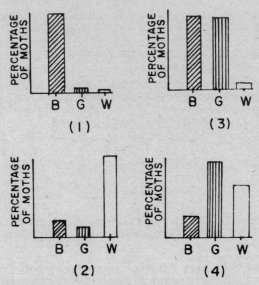

107. As a result of the fungus infection, the change in the moth wing color distribution would most probably occur by the
    (1) inheritance of an acquired characteristic
    (2) natural selection of favorable variations
    (3) ingestion of pigmentation from fungus spores
    (4) production of fungus-induced gene mutations

Directions: Base your answers to questions 108 through 110 on the information below and on your knowledge of the Hardy-Weinberg principle of gene frequencies.

In a certain population of rabbits, two alleles for coat color exist, tan and white. Tan is the dominant allele. The frequency of the gene for white coat color is .30. Foxes are also present in the area and rabbits constitute a major portion of their diet. Foxes recognize prey by the degree to which the prey contrast with their background.

108. What is the frequency of the gene for tan coat color?
    (1) .09   (3) .49
    (2) .21   (4) .70

109. In the entire population, the percentage of rabbits hybrid for tan coat color is
    (1) 42%   (3) 3%
    (2) 2%   (4) 49%

110. If the climate where the rabbits live were to change so that snow covered the ground for much of the year, what change might be expected in the rabbit population?
    (1) The frequency of the tan allele would increase.
    (2) The frequency of rabbits homozygous for the white allele would probably increase.
    (3) The frequency of rabbits homozygous for the tan coat color would increase.
    (4) The frequency of the white allele would decrease.

Group 6—Ecology

> If you choose this group, be sure to answer questions 111–120.

111. In which of the following biomes does most of the photosynthesis taking place on the Earth occur?
    (1) deciduous forests   (3) deserts
    (2) oceans   (4) coniferous forests

112. The rate of photosynthesis carried on by plants living in a body of water depends chiefly upon the
    (1) amount of molecular oxygen in the water
    (2) number of decomposers in the water
    (3) amount of light that penetrates the water
    (4) number of saprophytes in the water

113. Recent studies have found traces of the insecticide DDT accumulated in human fat tissue. A correct explanation for this accumulation is that
    (1) DDT is needed for proper metabolic functioning
    (2) DDT is passed along food chains
    (3) fat tissue absorbs DDT directly from the air
    (4) fat tissue cells secrete DDT

114. Which statement describes an ecological importance of insects?

(1) Insects are humans' chief competitor for available food.
(2) The destruction of all insects will maintain the balance of nature.
(3) The destruction of all insects would help maintain proper food webs.
(4) Insects are causing the bird population to decrease.

115. The most serious consequence of cutting down forests and overgrazing land is
(1) the prevention of flooding
(2) an increase in the chance of fire
(3) an increase in the number of predators
(4) the loss of topsoil cover

Directions (116–120): For each description in questions 116 through 120, select the biome, *chosen from the list below,* that is most closely associated with that description. [*A number may be used more than once or not at all.*]

Biomes
(1) Grassland
(2) Tundra
(3) Temperate deciduous forest
(4) Tropical rain forest
(5) Taiga
(6) Desert

116. This biome is found in the foothills of the Adirondack and Catskill Mountains of New York State, and supports the growth of dominant vegetation including maples, oaks, and beeches.

117. The characteristic climax vegetation in this biome consists of coniferous trees composed mainly of spruce and fir.

118. This biome receives less than 10 inches of rainfall per year. Extreme temperature variations exist throughout the area over a 24-hour period. Water-conserving plants such as cacti, sagebrush, and mesquite are found.

119. This biome receives the least amount of solar energy. The ground is permanently frozen (permafrost) throughout the year. During the summer season, plants quickly grow, reproduce, and form seeds during their short life cycle. Lichens and mosses grow abundantly on the surface of rocks.

120. A moderate, well-distributed supply of rain in this biome supports the growth of broad-leaved trees, which shed their leaves as winter approaches.

Answers

1. 2	25. 3	49. 4	73. 3	97. 2
2. 4	26. 2	50. 2	74. 3	98. 4
3. 1	27. 1	51. 3	75. 1	99. 1
4. 3	28. 2	52. 4	76. 2	100. 3
5. 4	29. 1	53. 4	77. 3	101. 4
6. 1	30. 4	54. 1	78. 2	102. 2
7. 2	31. 4	55. 4	79. 4	103. 1
8. 1	32. 3	56. 2	80. 2	104. 2
9. 2	33. 1	57. 1	81. 4	105. 1
10. 2	34. 4	58. 3	82. 1	106. 4
11. 3	35. 2	59. 4	83. 3	107. 2
12. 3	36. 3	60. 2	84. 1	108. 4
13. 4	37. 1	61. 2	85. 1	109. 1
14. 4	38. 3	62. 5	86. 3	110. 2
15. 3	39. 2	63. 1	87. 2	111. 2
16. 4	40. 1	64. 4	88. 4	112. 3
17. 2	41. 4	65. 4	89. 3	113. 2
18. 3	42. 2	66. 2	90. 2	114. 1
19. 1	43. 2	67. 1	91. 2	115. 4
20. 1	44. 3	68. 3	92. 4	116. 3
21. 3	45. 2	69. 3	93. 2	117. 5
22. 4	46. 1	70. 2	94. 3	118. 6
23. 2	47. 3	71. 2	95. 1	119. 2
24. 3	48. 1	72. 4	96. 3	120. 3

Explanatory Answers

1. **(2)** The idea of origination from preexisting life was developed through a series of experiments in abiogenesis and biogenesis. In the final analysis, Louis Pasteur developed the undeniable proof of biogenesis through his experimentation. INCORRECT CHOICES: (1) Not all living organisms require oxygen for respiration, e.g., yeast cells. (3) Again, the simple cells have no nervous system but do originate from prexisting life. (4) Some organisms carry on autotrophic nutrition.

2. **(4)** Regulation either by chemical or neural–chemical control is necessary to coordinate and maintain the organism's responses to its surroundings. INCORRECT CHOICES: (1) Nutrition is the process of ingestion and digestion culminating in the use of that food for energy and synthesis. (2) Respiration is the oxidation of glucose either anaerobically or aerobically to release energy. (3) Transport is the circulation of products of digestion and gases and waste products within the organism.

3. **(1)** The mitochondria are called the powerhouses of the cell and do produce great quantities of ATP. This process occurs during aerobic respiration. INCORRECT CHOICES: (2) The centrosomes play a role in animal cell reproduction. (3) The chloroplasts of plant cells produce glucose through photosynthesis. (4) The nuclei of the cell coordinate all of the cell processes.

4. **(3)** 1.5 millimeters, according to the conversion factor of 1 micron = 0.001 millimeters, equals 1500 microns. INCORRECT CHOICES: (1), (2), and (4) These equivalents do not compute using the formula above.

5. **(4)** The key to this question is the ratio of hydrogen to oxygen as 2:1. This ratio is characteristic only of the carbohydrates. INCORRECT CHOICES: (1) and (2) Lipids and fatty acids have a much greater ratio than 2:1. (3) The proteins contain nitrogen and do not have a 2:1 ratio of hydrogen to oxygen.

6. **(1)** Water is the universal solvent and therefore would play a major role in readying materials for their passage through the cell membrane. INCORRECT CHOICES: (2) Salt is a solute and water the solvent. It is the presence of water that makes the salt's diffusion possible. (3) and (4) Iron and phosphorus are soluble in water as was the salt. The statement then would be the same.

7. **(2)** The setae anchor the worm in its burrow. These structures dig into the side of the burrow so that backward pulling efforts are thwarted. INCORRECT CHOICES: (1) The earthworm does not possess chitinous appendages and an exoskeleton. (3) The earthworm does not possess ventral suckers or claws. (4) The earthworm does not possess mouth parts or scales.

8. **(1)** Both neurohumors and hormones play a major role in the regulation of body processess: the neurohumors with the transmission of the impulse, and hormones with most other body processes. INCORRECT CHOICES: (2) Chitin is a material composing the exoskeleton of some arthropods. (3) Urine is a liquid waste product excreted by humans. (4) Pigments are areas of coloration ranging from green through blue.

9. **(2)** The liver filters the blood and forms urea from the breakdown of amino acids. Water and carbon dioxide can result from respiration of liver cells. INCORRECT CHOICES: (1) Oxygen listed here is not a waste product. (3) Hormones listed in this choice are not waste products. (4) Neither hormones nor oxygen are waste products.

10. **(2)** Respiration for the grasshopper, a land dweller, is accomplished through a series of tracheal tubes leading to the interior of the body. The transport of oxygen into and carbon dioxide out of the cells occurs through these tubes. INCORRECT CHOICES: (1) Malpighian tubules in the grasshopper are used for the concentration of nitrogen wastes. (3) The contractile vacuoles of unicellular organisms excrete excess water. (4) The lungs of humans carry on the process of gas exchange.

11. **(3)** A lack of oxygen, arising, for example, as a result of strenuous exercise, leads to the anaerobic oxidation of glucose in muscle cells. The result is the formation of lactic acid and sore muscles. INCORRECT CHOICES: (1) Carbon dioxide is a by-product of respiration. (2) Nitrogen does not play a role in respiration. (4) Insufficient hydrogen does not cause the buildup of lactic acid.

12. **(3)** Energy necessary to convert ADP to ATP must come from other energy molecules. These molecules then must be organic in nature. INCORRECT CHOICES: (1) Hormones regulate body processes. (2) Sunlight activates the chlorophyll molecules in plant cells. (4) Inorganic molecules do not have in their bonding enough energy to form ATP.

13. **(4)** The earthworm, a land dweller, has adapted its transport system to form a closed system. INCORRECT CHOICES: (1), (2), and (3) All of these choices are water-dwellers and rely on their environment (water) to provide all materials and facilitate diffusion (transport).

14. **(4)** ATP is a compound in which energy is stored temporarily. To retain a higher concentration of salts within the Paramecium body than in the external environment, energy must be used. That energy is stored in ATP. If energy was expended, the natural flow of materials would be to the outside of the cell. INCORRECT CHOICES: (1) Pinocytosis is the engulfment of large molecules of food particles and water by the formation of pinocytic vesicles. Pinocytosis requires the expenditure of energy. (2) Cyclosis is the streaming motion of the cytoplasm. (3) CO_2 is used as a raw material for photosynthesis and as a waste product of respiration.

15. **(3)** Mechanical digestion results in a physical change. No hydrolysis of the meat would occur. The result would be smaller pieces of meat.

INCORRECT CHOICES: (1), (2), and (4) The question is referring to mechanical digestion and these answers are for chemical digestion.

16. **(4)** The complete hydrolysis of carbohydrates results in the formation of the simple sugars such as glucose. INCORRECT CHOICES: (1) Glycogen is animal starch. (2) Urea molecules are waste products formed from the breakdown of amino acids. (3) Carbon dioxide is a gas released as a waste product from respiration.

17. **(2)** The answer of Ameba and Paramecium is correct because these organisms are unicellular and do not possess a sophisticated digestive system. INCORRECT CHOICES: (1), (3), and (4) All of these choices include multicellular organisms that possess a special digestive system.

18. **(3)** Cilia are locomotor organelles in the Paramecium. INCORRECT CHOICES: (1) Protozoa do not possess muscles. (2) Tentacles are found in the Hydra. (4) Setae are used in locomotion in the earthworm.

19. **(1)** The grasshopper, an insect, possesses the ventral nerve cord. INCORRECT CHOICES: (2) The Ameba does not possess a ventral nerve cord or any nerve cord for that matter. (3) The Hydra possesses a nerve net. (4) Humans have a dorsal nerve cord.

20. **(1)** The nervous system regulates the impulses initiated by environmental stimuli and the endocrine system secretes hormones that regulate the general body metabolism. INCORRECT CHOICES: (2) The excretory system listed here is involved only with the removal of metabolic wastes. (3) The locomotive system here is only involved with the movement of the organism. (4) Both of these choices listed in this answer have just been discussed.

21. **(3)** Complex proteins can only be made through the process of synthesis. INCORRECT CHOICES: (1) Oxidation is the loss of hydrogen atoms. (2) Digestion is the chemical or mechanical breakdown of food. (4) Enzymatic hydrolysis is the digestion of a substance controlled by enzymes as catalysts.

22. **(4)** In that the root is underground in many plant species, it would stand to reason that this part of the plant would anchor the species. INCORRECT CHOICES: (1) The leaf is specialized for photosynthesis. (2) The flower is specialized for sexual reproduction. (3) The stem is specialized for transport and some storage.

23. **(2)** Auxins are plant hormones found at the meristematic ends of leafs, stems, and roots. These chemicals stimulate growth. Therefore, by removing the stem of a plant, you effectively remove the auxins. INCORRECT CHOICES: (1) Sugars would be found in the palisade cell layer of the internal leaf and in the roots. (3) Carbon dioxide would be found escaping through the stomates and also entering through the stomates. (4) Oxygen would be found as a waste gas escaping through the stomates.

24. **(3)** Oil as an organic compound is synthesized from simpler elements and compounds, namely, glycerol and fatty acids. INCORRECT CHOICES: (1) The only thing absorbed from the sun is energy. (2) The only substances absorbed by the roots from the soil are water and minerals. (4) Chlorophyll molecules synthesize glucose from the sun, as well as water, enzymes, and carbon dioxide.

25. **(3)** Nitrogenous compounds contain nitrogen, the key element in proteins. We can then assume that these nitrogenous compounds release nitrogen for the synthesis of proteins. INCORRECT CHOICES: (1) Glucose does not contain nitrogen in its structure. (2) Waxes belong to the fat category, and as such are composed of glycerol and fatty acids. (4) Starch molecules are polymers of carbohydrates.

26. **(2)** Photosynthesis gives off oxygen as a waste product. It is then released via the stomates into the atmosphere. INCORRECT CHOICES: (1) Aerobic respiration uses oxygen as a reactant rather than producing it as a product. (3) Transpiration is concerned with the removal of water through the stomates. (4) Capillary action serves to move water from the roots to the leaves.

27. **(1)** Phloem vessels are specialized to transport food produced in the leaf to the roots for storage. INCORRECT CHOICES: (2) Xylem tubes are specialized for transport of water and minerals from the root to the leaf. (3) Root hairs increase the absorptive surface of roots. (4) Chloroplasts are necessary for the manufacture of glucose.

28. **(2)** Maple trees as well as other land plants, have no specialized digestive systems and rely on vacuoles, which are rather large, to hydrolyze their food. Since this process occurs within the cell, it is called intracellular digestion. INCORRECT CHOICES: (1) Plants such as the maple tree, do not perform extracellular digestion. (3) Only intracellular digestion occurs. (4) If the plant is to survive, it must carry on some form of digestion.

29. **(1)** The term heterotroph means that the organism relies on preformed organic compounds in other organisms to supply its nutritive require-

ments. INCORRECT CHOICES: (2) Autotrophic organisms need to live in a sunny environment, heterotrophs do not. (3) Most heterotrophs are mobile only in that the mobility furnishes a wider range of preformed organics. (4) This choice is a description of an autotroph.

30. **(4)** Blue light will cause a faster rate of photosynthesis because it is not totally reflected as is the green light. (Remember that the leaf is green only because it is reflecting green light.) INCORRECT CHOICES: (1) Photosynthesis will occur in the plant under the blue light. (2) Photosynthesis will not occur in the green light according to the answer above. (3) The plant under green light will not undergo photosynthesis at a faster rate.

31. **(4)** Vegetative propagation involves the roots, stems, or leaves. As pictured in the diagram, we see new plants arising from an above ground stem. INCORRECT CHOICES: (1) Cleavage is the rapid division of cells in the early development of the embryo. (2) Fission is the splitting of the cytoplasm and the exact division of the nuclear contents to form two new cells. (3) Zygote formation implies sexual reproduction as an end to fertilization.

32. **(3)** The word gamete or sex cell implies that the chromosomal number is monoploid. This means that one half the normal amount of chromosomes is present. Therefore, one member of each pair of homologues would appear in each gamete. INCORRECT CHOICES: (1) A zygote is diploid and the result of fertilization. (2) A multicellular embryo is an embryo composed of more than one cell and is diploid. (4) A cheek cell is a somatic cell and is diploid.

33. **(1)** The anther is the male reproductive structure of the flower and it would be here that pollen formation (sperm nuclei) would occur. INCORRECT CHOICES: (2) The stigma is part of the female reproductive structure. (3) The pistil is also a part of the female reproductive structure. (4) The ovary is the site of fertilization and contains the ovule.

34. **(4)** The primary spermatocyte divides mitotically to form four monoploid sperm cells. This process takes place in two steps and results in four gametes. INCORRECT CHOICES: (1), (2), and (3) According to the outline above these numbers could not be possible.

35. **(2)** We see in the diagram the diploid adult forming monoploid gametes. This process is meiosis. We then see cleavage occurring but without first seeing fertilization. This process would be called parthenogenesis. INCORRECT CHOICES: (1) Meiosis is correct but budding is incorrect. (3)

There appears no fertilization in the diagram. (4) Again fertilization is omitted in the diagram in that this process is parthenogenetic.

36. **(3)** As the chromosomes in interphase replicate, the pair of homologues are called chromatids. INCORRECT CHOICES: (1) A centromere is an oval body which connects the two chromatids. (2) A homologue is a single chromosome of an identical pair of chromosomes. (4) A tetrad is formed when a pair of homologues replicate, such as we find in meiosis.

37. **(1)** Plant cells divide their cytoplasm by the formation of a cell plate. Animal cells do not possess this structure. INCORRECT CHOICES: (2) The centromere connects two chromatids. (3) Chromosome replication occurs in both plant and animal cells. (4) Centrioles are found only in animal cells.

38. **(3)** Mitosis provides for the exact duplication of hereditary material and subsequent division of that material to two daughter cells. This provides for the exact replication of two cells, genetically the same as the parent. INCORRECT CHOICES: (1), (2), and (4) These choices would not apply because they do not provide for the exact duplication of genetic material.

39. **(2)** Hermaphroditism is characterized by both sexes appearing in the same body, as we see in the earthworm or Hydra. INCORRECT CHOICES: (1) Hybrids contain two different alleles for a given trait. (3) Phagocytes are cells which attack and engulf invading antigens. (4) Parasites are organisms that feed on other organisms.

40. **(1)** The nucleotide is the basic component of a nucleic acid such as DNA or RNA. INCORRECT CHOICES: (2) RNA does not exist as a double helix. (3) Deoxyribose sugar is a component of DNA. (4) Thymine is a nitrogen base found only in DNA.

41. **(4)** Grafting is the attachment of a scion to the stock of a tree. This method allows for the propagation of several species on one tree. INCORRECT CHOICES: (1) Hybridization is the form of controlled breeding. (2) Pollination is the transfer of pollen from the anther to the stigma. (3) Fertilization is the union of egg and sperm cells.

42. **(2)** The genetic composition of skin cells is such that certain pigment-containing organelles containing melanin can be influenced by sun exposure to produce more than the normal amount of coloration. This external influence is an example of the environment affecting genetic control. INCORRECT CHOICES: (1) The tanning of the skin is not considered

to be a mutation. (3) The theory of acquired characteristics were disproven by Weismann. (4) Albinism is recessive but it has nothing to do with the effect of sunlight on the rate of tanning.

43. **(2)** If mutations were beneficial to the organism, we would find more mutations occurring than we do. However, the rate of mutations is so low that we say they are recessive and disadvantageous to the cell. INCORRECT CHOICES: (1), (3), and (4) Following the statement from above, we can see why these choices would be incorrect.

44. **(3)** Nondisjunction, as the name implies, is the failure of something to fall away from or disjoin. This applies to chromosomes and results in an extra chromosome. INCORRECT CHOICES: (1) Crossing-over occurs in the tetrad stage of meiotic division. (2) Polyploidy is the condition in which extra sets of chromosomes exist beyond the diploid number. (4) Recombination is the recombination of linkage groups after crossing over has occurred.

45. **(2)** If we set up a Punnett square using the following: $P_1 = Ai \times Bi$, we end up with a set of blood types which include, AB, Ai, Bi, ii. Since blood types A,B,O have already been accounted for, there remains blood type AB which is 25% of the square. INCORRECT CHOICES: (1), (3), and (4) These choices do not appear logical using the information from above.

46. **(1)** Using the Punnett square we find that this cross $(XcXc \times XY)$ produced two carrier females (XcX) and two colorblind males (XcY). No colorblind females appear. INCORRECT CHOICES: (2), (3), and (4) These choices do not appear logical using the information above.

47. **(3)** In a cross of contrasting traits, the trait that appears in the first filial generation is dominant. This is Mendel's law. INCORRECT CHOICES: (1) The law of independent assortment states that the separation of gene pairs on a pair of chromosomes is entirely independent of the assortment of other gene pairs on other chromosome pairs. (2) The law of segregation states that a pair of genes may be segregated during the formation of gametes. (4) Incomplete dominance states that neither gene is dominant or recessive but there exists a blending of the traits.

48. **(1)** Evolutionary relationships allow us to compare structure, embryo formation, biochemistry, and homologous organs. This in turn allows us to make inferences concerning the placement of species in the classification scheme. INCORRECT CHOICES: (2) Similar species may not live in similar habitats. (3) Geographic distribution is based on many factors

not just on similarity. (4) Mendelian principles relate to the heredity of the individual not the classification of that species.

49. **(4)** This particular cross is Mendel's law of dominance. We see the appearance of all dominant offspring from a mating of a dominant parent and a recessive parent. This leads us to conclude that the parents were homozygous dominant and recessive. INCORRECT CHOICES: (1), (2), and (3) These choices are incorrect if you follow the logic from above.

50. **(2)** The lichens will begin to turn the bare rock into tiny soil particles which will then lead to the introduction of rooted species. INCORRECT CHOICES: (1) If they were climax organisms they would be very stable and dominant to other species. (3) A primary consumer is one which feeds on autotrophs. Lichens are autotrophic. (4) Decomposers feed on the dead or decaying remains of organics, and lichens are autotrophs.

51. **(3)** Since the pig does not use the two toes, Lamarck would have said that the pig would lose the toes. INCORRECT CHOICES: (1) Natural selection was a theory proposed by Darwin. (2) The theory of mutations was developed by DeVries after Darwin proposed his ideas. (4) Synapsis is the pairing of homologues during meiosis.

52. **(4)** The appearance of variations was explained differently by Lamarck, Darwin, and finally Weismann. It wasn't until the idea of mutations was made clear that evolutionary trends became clear. INCORRECT CHOICES: (1) Differentiation in embryonic development has nothing to do with explaining mutations. (2) Stability of gene pools does not allow for the appearance of mutations. (3) The extinction of dinosaurs was due to their failure to adapt quickly enough to a rapidly changing environment.

53. **(4)** Comparative biochemistry allows biologists to determine if the organisms have a common ancestory. INCORRECT CHOICES: (1) This is true of all enzymes in all species. It does not tell us if there is a related thread between two organisms. (2) This is also true but does not explain the relationship. (3) This statement is not true.

54. **(1)** If the rock layers are undisturbed then we could find the oldest fossils at the bottom and the youngest at the top. INCORRECT CHOICES: (2) This statement could not be true if we follow the idea of sedimentation occurring and slowly building up the earth layer. (3) Not all fossils of species would show homologous structures because their environments would be different. (4) Again this would be entirely impossible.

55. **(4)** The description given in the question defines the feeding habits as those of scavengers and predators. INCORRECT CHOICES: (1), (2), and (3) These choices do not meet the descriptions given.

56. **(2)** Most producers, with the exception of chemosynthetic bacteria, contain chlorophyll which allows them to utilize radiant energy to produce glucose. INCORRECT CHOICES: (1) Parasites feed on a host organism and would then be heterotrophic. (3) Carnivores eat meat not plants. (4) Most producers do not liberate nitrogen.

57. **(1)** Abiotic means nonliving. The only choice here that fits this definition is water. INCORRECT CHOICES: (2), (3), and (4) All of these choices are biotic.

58. **(3)** The transformation stated in the question is photosynthesis. The only choice that fulfills the requirements is the rosebush. INCORRECT CHOICES: (1), (2), and (4) All of these choices are heterotrophic.

59. **(4)** We would assume from the question that the snake feeds on the other organisms listed. If the number of snakes greatly overshadowed the numbers of the other species, those species would die out and so too would the snake. INCORRECT CHOICES: (1), (2), and (3) In looking at these choices we could not say that any of them ate the snake. Therefore we would assume that the snake would eat them.

60. **(2)** Aphids eat plants and are therefore herbivores. INCORRECT CHOICES: (1) The rosebush is a producer. (3) and (4) These choices are all carnivorous.

61. **(2)** This is a fatty acid. Fats or lipids are composed of three fatty acids and one glycerol molecule. INCORRECT CHOICES: (1) This represents an amino acid, characterized by an amino (or $-NH_2$) and carboxyl or acid ($-COOH$ group) surrounding a central carbon atom. Also bonded to this carbon is an R variable group and a hydrogen. (3) This is the structural formula for carbon dioxide. (4) This represents a glucose molecule. (5) This depicts the disaccharide, maltose.

62. **(5)** Maltose, a double sugar, is formed when two glucose molecules unite by splitting out one molecule of water. INCORRECT CHOICES: See number 61 for (1), (3), and (4). (2) This represents a fatty acid which has a long chain of carbon and hydrogen atoms with an acid (or $-COOH$) ending.

63. **(1)** All proteins are polymers or chains of amino acids. INCORRECT CHOICES: See numbers 61 and 62.

64. **(4)** Glucose, $C_6H_{12}O_6$, is a simple sugar or monosaccharide. INCORRECT CHOICES: See numbers 61 and 62.

65. **(4)** PGAL is a three-carbon organic compound which is the first product of photosynthesis. From PGAL, plants can make all other organic compounds including glucose. INCORRECT CHOICES: (1) Chlorophyll is needed to capture light energy for photosynthesis. (2) Oxygen is another product of photosynthesis. (3) Water is a raw material required for photosynthesis.

66. **(2)** Lipases are enzymes which control the breakdown of lipids, (fats or oils). INCORRECT CHOICES: (1) Proteases are enzymes which act on proteins. (3) Maltases control reactions involving the double sugar, maltose. (4) Vitamins frequently function as the coenzyme part of enzyme molecules.

67. **(1)** Ethyl alcohol and carbon dioxide are products of the form of anaerobic respiration called fermentation. INCORRECT CHOICES: (2) Complex sugars are formed by the joining of simple sugars. (3) Oxygen is a product of photosynthesis. (4) Starches are formed by the union of many simple sugars by dehydration synthesis.

68. **(3)** When the enzyme concentration reaches 0.3 mg/ml, the rate of enzyme action levels off because all of the fixed amount of substrate is being acted upon by the enzymes. INCORRECT CHOICES: (1), (2), and (4) The concentrations are all below 0.3 mg/ml and the maximum rate of reaction has not been reached. These enzyme concentrations are not high enough to use all the available substrate.

69. **(3)** The rate of enzyme action remains the same because all of the fixed amount of substrate available has been used. INCORRECT CHOICES: (1) The rate of reaction would decrease only if the amount of available substrate, which is fixed, decreased. (2) The rate of reaction would increase only if the amount of substrate, which is fixed, increased.

70. **(2)** The rate of reaction would increase because it is limited to a maximum of 0.3 mg/ml by the constant level of substrate concentration. INCORRECT CHOICES: (1) The rate of reaction would decrease if substrate were *removed* not added. (3) The rate of reaction only remains the same if the level of substrate is constant.

71. **(2)** Blood in chambers 2 and 4 has returned from the lungs and is rich in oxygen. INCORRECT CHOICES: (1) Blood in chamber 1, the right atrium, has returned from the body and is poor in oxygen. (3) Blood in chamber 3, the right ventricle, will be pumped to the lungs where it will pick up oxygen. (4) Blood in the right side of the heart, chambers 1 and 3, is poor in oxygen.

72. **(4)** Blood in *D*, the vena cava, is returning from the body, and blood in *B*, the pulmonary arteries, is going to the lungs. Both are rich in CO_2 and poor in O_2. INCORRECT CHOICES: (1) *A* is the aorta, the largest artery in the body. Arteries usually carry blood rich in oxygen. (2) *C*, the pulmonary vein, returns blood rich in oxygen to the heart. (3) *C* and *A* are both carrying blood which has passed through the lungs and is rich in oxygen.

73. **(3)** The alveoli are air sacs located in the lungs, structures 5 and 6. INCORRECT CHOICES: (1) 1 and 3 are chambers of the heart. (2) and (4) *A*, *B*, *C*, and *D* are blood vessels which do not contain alveoli.

74. **(3)** Impulses originate in a receptor and pass through sensory, associative, and motor neurons to the effector. INCORRECT CHOICES: (1) A reflex arc does not pass from the effector, a muscle or gland, to a receptor, a sense organ. It travels in the opposite direction. (2) The impulse from a receptor travels through a sensory not a motor neuron; a motor neuron carries an impulse to an effector. (4) A receptor, not an effector, receives stimuli; an effector is innervated by a motor neuron.

75. **(1)** The pancreas secretes insulin which controls the level of sugar in the blood. INCORRECT CHOICES: (2) The parathyroid controls calcium metabolism. (3) The salivary glands secrete saliva, which contains enzymes for starch digestion. (4) The thyroid gland controls the rate of cellular metabolism.

76. **(2)** Ligaments connect bones to bones. INCORRECT CHOICES: (1) Tendons attach muscles to bones. (3) Muscles are tissues specialized for movement. (4) Neurons or nerve cells are adapted to carry an electrochemical message or impulse.

77. **(3)** Fecal wastes are formed in the large intestine, number 3. INCORRECT CHOICES: (1) This is the esophagus. (2) This indicates the stomach. (4) This points to the small intestine.

78. **(2)** Proteins are first digested in the stomach. INCORRECT CHOICES: See number 77.

79. **(4)** All lipid digestion starts and is completed in the small intestine. INCORRECT CHOICES: See number 77.

80. **(2)** Gastric glands, located in the walls of the stomach, secrete juices needed for protein digestion. INCORRECT CHOICES: See number 77.

81. **(4)** In both the monocot and dicot seeds, the cotyledons store food, in the form of starch, for the developing embryo. INCORRECT CHOICES: (1) Flowers develop from stems not seeds. (2) Stems develop from the epicotyl of the seeds. (3) Roots develop from the hypocotyl of the seeds.

82. **(1)** Oxygen is needed by the maple seed for respiration. INCORRECT CHOICES: (2) Seeds can germinate in media other than soil. (3) Seeds do not carry on photosynthesis; therefore, they do not need chlorophyll. (4) Seeds can germinate in the presence of light.

83. **(3)** Scrotal temperature is 2–4 degrees lower than body temperature; this lower temperature is optimal for sperm production. INCORRECT CHOICES: (1) The body cavity would offer greater protection for the testes. (2) The testes are not in contact with the atmosphere; oxygen travels to the testes through the blood. (4) A duct, the vas deferens, carries sperm from the testes to the prostate and urethra.

84. **(1)** Differentiation is the formation of specialized tissues and organs from the three embryonic layers which develop as a result of gastrulation. INCORRECT CHOICES: (2) Menstruation, the periodic shedding of the lining of the uterus, takes place if fertilization does not occur. (3) Ovulation is the release of the egg from the ovary. (4) Fertilization is the union of the egg and sperm to form the embryo.

85. **(1)** The amnion is a fluid-filled sac which surrounds the developing chick embryo.

86. **(3)** The chorion is a thin membrane found directly under the shell; its function is to provide a thin, moist membrane for gas exchange.

87. **(2)** The allantois is a sac connected to the embryo; its functions include excretion and respiration. INCORRECT CHOICES: (4) The yolk sac surrounds the stored food in the yolk; blood vessels penetrate the yolk sac and carry food back to the developing embryo.

88. **(4)** D represents the uterus in which the embryo implants. INCORRECT CHOICES: (1) The egg develops in A, the ovary. (2) B indicates the release of the egg from the ovary. (3) C represents the oviduct.

89. **(3)** Fertilization, the union of egg and sperm, normally takes place in the oviduct. INCORRECT CHOICES: (1) The egg is not fertilized until after it leaves the ovary, A. (2) B depicts the release of the egg prior to fertilization. (4) The egg will disintegrate if it is not fertilized before it reaches D.

90. **(2)** Eggs are depicted leaving the follicle of the ovary by *B*; this is ovulation. INCORRECT CHOICES: (1) The ovary, *A*, produces eggs by meiosis in tiny cavities called follicles. (3) The oviduct, *C*, transports the eggs to the upper end of the uterus. (4) After ovulation and fertilization, the developing embryo implants in the uterus, *D*.

91. **(2)** A gene mutation is any change that affects the base sequence in an organism's DNA; this can be an addition, deletion, or substitution. INCORRECT CHOICES: (1) Replication is the production of a duplicate DNA molecule. (3) A chromosome mutation is a change in the chromosome structure or in the numbers of chromosomes. (4) Independent assortment indicates that genes for two different traits that are not located on the same chromosome are inherited independently.

92. **(4)** Weak hydrogen bonds connect the nitrogen bases, these bases connect the two strands of the DNA molecule. INCORRECT CHOICES: (1) Phosphate and sugar molecules form the sides of the DNA strands. (2) Deoxyribose, not ribose, is found in DNA. (3) DNA is the deoxyribose molecule.

93. **(2)** In DNA, the adenine base pairs with thymine. INCORRECT CHOICES: (1) Only in RNA does adenine pair with uracil. (3) Cytosine pairs with guanine. (4) In DNA guanine pairs with cytosine; adenine with thymine.

94. **(3)** Genes have been found in two self-duplicating structures, the mitochondria and chloroplasts. INCORRECT CHOICES: (1) Food vacuoles contain enzymes for digestion. (2) Contractile vacuoles store and excrete excess water. (4) Cell walls, the nonliving outer boundary of plant cells, are produced under the direction of genes found in the nucleus.

95. **(1)** Transfer RNA carries amino acids to the ribosomes; the amino acids are linked to form proteins according to instructions coded in the messenger RNA. INCORRECT CHOICES: (2) Proteins are synthesized in the cytoplasm from amino acids. (3) Transfer RNA molecules are made from instructions coded in the DNA. (4) DNA provides a template for the synthesis of messenger RNA.

96. **(3)** Uracil is a nitrogen base which is found only in RNA. INCORRECT CHOICES: (1), (2), and (4) Guanine, cytosine, and adenine are nitrogen bases found in both DNA and RNA.

97. **(2)** Messenger RNA is synthesized in the nucleus; it transports the base sequence code from the DNA to the ribosomes. INCORRECT CHOICES: (1) The sequence of transfer RNA nucleotides at the ribosome is determined by the messenger RNA sequence and not the reverse. (3) The amino

acid sequence is determined by the sequence of nucleotides in messenger RNA, not vice versa. (4) DNA serves as the template for the manufacture of RNA.

98. **(4)** Using the triple code given and reading the portion of messenger RNA depicted, G-C-U codes alanine, U-U-U phenylalanine, and G-G-U glycine. INCORRECT CHOICES: (1) The amino acids are not in the correct sequence for the messenger RNA depicted. (2) U-U-U is the code for phenylalanine not glycine. (3) The sequence of the amino acids does not correspond to that called for by the messenger RNA.

99. **(1)** Ribosomes are found in the cytoplasm. INCORRECT CHOICES: (2) Centrioles, not ribosomes, are found in the area known as the centrosome. (3) RNA is stored in the nucleolus in the nucleus. (4) Messenger RNA is synthesized in the nucleus and then migrates to the ribosomes in the cytoplasm.

100. **(3)** Polypeptides are chains of amino acids which are synthesized at the ribosomes. INCORRECT CHOICES: (1), (2), and (4) Fats, vitamins, and carbohydrates are synthesized under the direction of enzymes which are not located on the ribosomes.

101. **(4)** According to the modern theory of evolution, species are undergoing change due to natural selection. INCORRECT CHOICES: (1) The genetic material is DNA, a nucleic acid. (2) Similar enzyme-controlled reactions are found in all cells. (3) With the exception of organisms such as viruses and slime molds, most organisms are composed of a cell or cells.

102. **(2)** The first organisms were heterotrophs and took in preformed organic compounds. Oxygen was not available for respiration; they were anaerobes. INCORRECT CHOICES: (1) Aerobes require oxygen which was not present in the primitive atmosphere. (3) See explanation for incorrect choices 1 and 4. (4) Autotrophs can synthesize organic compounds from inorganic; the required carbon dioxide was not present in the primitive atmosphere.

103. **(1)** The gene pool is the sum of all the genes collectively present in a population. INCORRECT CHOICES: (2) Genotype is a description of the genes for a particular trait present in an individual. (3) Frequency is the percentage of genes in the population for a specific allele. (4) Phenotype is the appearance of an individual with respect to a particular trait.

104. **(2)** Barriers which prevent successful interbreeding between organisms living at the same time in the same area result in reproductive isolation.

INCORRECT CHOICES: (1) Artificial selection occurs when organisms with particular traits are selected for breeding. (3) Survival of the fittest is Darwin's theory which states that those best adapted survive and pass on their traits. (4) Overproduction is a description of the phenomenon that more offspring are produced than can possibly survive.

105. **(1)** According to the Hardy-Weinberg principle, under the following conditions the gene frequencies in a population remain constant: large population, random mating, no migrations, and no mutations. INCORRECT CHOICES: (2), (3), and (4) Under the conditions described there will be a stable gene pool and no change in gene frequency. All other answers indicate a change in gene frequency.

106. **(4)** Moths with a gray-white color would be best suited to the changed environment; they would survive and pass on their traits for wing color. The frequency of these genes would then increase in the population. INCORRECT CHOICES: (1) Brown colored moths would lack protective coloration and decrease in numbers. (2) Both gray and white colored moths would have the advantage of protective coloration. (3) When the tree bark changed color, brown moths would decrease in number and white moths would increase.

107. **(2)** Natural selection states that individuals that survive are the ones best fitted to exist in their environment, i.e., those with protective coloration. INCORRECT CHOICES: (1) Modern theory does not accept the concept of inheritance of acquired characteristics since only changes involving sex cells can be inherited. (3) Ingestion of pigment would not change the genes for wing color and therefore would not be inherited. (4) Fungus-induced mutations would not necessarily result in a change in wing coloration.

108. **(4)** q = frequency of the recessive allele for white coat = 0.30
p = frequency of the dominant allele for tan coat
$p + q$ = 1.0 or 100%
If $q = 0.30$, then $p = 0.70$ INCORRECT CHOICES: (1), (2), and (3) These are incorrect according to the calculations above.

109. **(1)** If $p = 0.70$ and $q = 0.30$, according to Hardy-Weinberg principle, the percentage of individuals with the hybrid trait is $2pq$. $2pq = 2 (0.70 \times 0.30) = 2 (0.21) = 0.42 = 42\%$ INCORRECT CHOICES: (2) and (3) According to the calculations above, these are incorrect answers. (4) The frequency of the homozygous or pure dominate = $p^2 = (0.7)^2 = 0.49 = 49\%$.

110. **(2)** Rabbits with white coat color would show the least contrast with their background. White coat color is due to homozygous recessive alleles. INCORRECT CHOICES: (1) Tan alleles result in tan coat color which would increase the changes of foxes recognizing them as prey. (3) Rabbits homozygous for tan coat color would be less likely to survive and therefore the frequency of the allele for tan coat would decrease. (4) Since the white color is selected for in the new environment, the gene or allele for white will increase not decrease.

111. **(2)** About 80% of the total photosynthesis on earth takes place in the waters. More than 70% of the earth's surface is covered by water. INCORRECT CHOICES: (1) Most photosynthesis is carried on by unicellular algae which are not abundant in deciduous forests. (3) The water supply for photosynthesis is limited in the desert. (4) The amount of sunlight available for photosynthesis is limited in coniferous forests.

112. **(3)** Although water absorbs much light energy, most photosynthesis takes place near the surface of the oceans because the deeper regions are too dark. Light intensity varies greatly with depth from the surface. INCORRECT CHOICES: (1) Although available oxygen is a factor in the rate of photosynthesis, there is sufficient oxygen dissolved in the surface of the oceans. (2) Decomposers do not produce the raw materials for photosynthesis. (4) Saprophytes live on dead or decaying material and do not carry on photosynthesis.

113. **(2)** DDT is a long-lasting insecticide which has entered the food chain upon which human beings depend. INCORRECT CHOICES: (1) DDT is a human-made insecticide not normally found in or needed by humans. (3) Fat tissue absorbs DDT from the blood. (4) Fat cells store DDT which is ingested along with the food.

114. **(1)** Since insects are human beings' chief competitors for food, their numbers are an important ecological factor. INCORRECT CHOICES: (2) The destruction of all insects would destroy the balance of nature since many animals such as birds, prey on insects. (3) The destruction of insects would destroy needed food for other animals. (4) Insects provide food for birds; they increase the bird population.

115. **(4)** Forests and soil cover plantings hold valuable topsoil and prevent erosion. Loss of them results in erosion of topsoil. INCORRECT CHOICES: (1) Cutting down trees increases, not decreases, the chances of flooding. Removal of soil and forest cover has led to disastrous floods. (2) Cutting down trees would reduce forest fires. (3) Removal of trees and ground cover would disrupt the food chain by eliminating the producers.

116 through 120 will be treated as a single unit to avoid repetition.

116. **(3)** In a temperate deciduous forest, the winters are cold, the summers warm, and the rainfall moderate; these conditions are found in the Adirondack and Catskill Mountains. Maples, oaks, and beeches are deciduous trees and lose their leaves in the winter.

117. **(5)** The taiga is a coniferous forest with pine or cone-bearing trees such as spruce and fir.

118. **(6)** The desert receives very little rainfall; the days are very hot and the temperature drops drastically at night. Only small, water-conserving plants can survive.

119. **(2)** The tundra, located at the poles, is characterized by very low temperatures, frozen subsoil, and limited vegetation. Only small plants with a limited growing season can survive.

120. **(3)** In a temperate deciduous forest there is sufficient rainfall to support the growth of large, deciduous trees which lose their leaves in the winter. INCORRECT CHOICES: (1) In a grassland, there is insufficient rainfall to support large trees; the winters are cold and long, the summers hot and short. (4) Tropical rain forests are located near the equator at sea level and are areas of high temperature and humidity which support luxuriant growth.

June 1982 Examination

Topic Analysis Key

If you had these answers wrong	Read these Chapters
1, 3, 4, 5, 6, 14, 16, 61, 62, 63, 64, 66, 68, 69, 70	The Story of Life
2, 7, 8, 9, 10, 11, 12, 13, 15, 17, 18, 19, 20, 21, 29, 71, 72, 73, 74, 75, 76, 77, 78, 79, 80	Maintenance in Animals
22, 23, 24, 25, 26, 27, 28, 30, 65, 67	Maintenance in Plants
31, 32, 33, 34, 35, 36, 37, 38, 39, 41, 81, 82, 83, 84, 85, 86, 87, 88, 89, 90	Reproduction and Development
40, 42, 43, 44, 45, 46, 47, 49, 52, 91, 92, 93, 94, 95, 96, 97, 98, 99, 100	Genetics
48, 51, 53, 54, 101, 102, 103, 104, 105, 106, 107, 108, 109, 110	Evolution and Diversity
50, 55, 56, 57, 58, 59, 60, 111, 112, 113, 114, 115, 116, 117, 118, 119, 120	Plants and Animals in Their Environment